Six-Step HVAC Maintenance Recovery

(A step-by-step guide to energy optimization, comfort improvement and indoor air quality.)

Tom Olson

Copyright © 2014 Tom Olson
All rights reserved.

ISBN-13:
978-1502764980

ISBN-10:
1502764989

CONTENTS

Foreword 6
Preface   7
Introduction   10

Chapter 1 – Fix what's broken

1.A – Temperature control air compressor 14
1.B – Steam Traps   21
1.C – Air-handling units   28
1.D – Control valves   37
1.E – Boilers   43

Chapter 2 – Clean what's dirty

2.A – Bird screens.. 50
2.B – Unit vents   55
2.C – Unit vent filters   59
2.D – Air-handling units   62
2.E – Air-handling unit filters   67
2.F – Condenser coils   71
2.G – Boilers   78
2.H – Boiler room and equipment rooms   81

Chapter 3 – Changes in operations made possible by Steps 1 & 2

3.A – Boilers... 85
3.B – Steam pressure   89
3.C – Hot water reset   91
3.D – Night setback   92
3.E – Lighting   95
3.F – Electric boost heaters   97
3.G – Summer water heaters   99
3.H – Timers and time clocks   102
3.I – Ultrasonic fault detectors   105

Chapter 4 – Revise temperature control sequences

4.A – Eliminate simultaneous cooling and heating 109
4.B – Enthalpy control   117
4.C.- Natatorium (swimming pool) air-handling units   119

Chapter 5 – Installation of new technology hardware

5.A – Best Value selection 127
5.B – Boilers   128
5.C – Improved hot water reset   141
5.D – Building automation systems (BAS)   150
5.E – Variable frequency drives (VFD's)   153
5.F – Swimming pool heaters   155
5.G – Swimming pool blankets   157
5.H – Lighting   158

Chapter 6 – Preventive maintenance

6.A – Training 159
6.B – Planning   162
6.C – Scheduling   166
6.D – Benchmarking   168
6.E – Commissioning   171
6.F – Do it

Chapter 7 - BONUS - More things that are good to know, but don't fit into the six-steps

7.A – Best Practices 174
7.B – Performance contracting   175
7.C – Carbon dioxide (CO2)   177
7.D – Fuel cost comparison   179
7.E – Contact   181

Foreword

I have worked in commercial facilities for 40 years and have known Tom Olson for 30 of those years. At the time I first met Tom I was the new Buildings and Grounds Supervisor at the Hopkins Schools, a suburb of Minneapolis, MN. Tom's company, Climate Makers, was doing some service and pneumatic control work for my new district. My assistant, Tom Robinson, told me he was very happy with the service Tom's company was providing.

I set up a meeting with Tom to learn what services were being provided and where he thought he could help us reduce our heating and air-conditioning energy consumption, without decreasing classroom and office air quality. We talked a while and then took a tour of three of our school buildings. I soon realized that I was with a person who was truly knowledgeable about HVAC operations and was very passionate about reducing energy consumption and increasing building comfort and air quality.

During the next two hours I learned a lot about the first two steps of Tom's program: Fix What's Broken and Clean What's Dirty. Over the next years, with Climate Makers, and a newly educated maintenance staff, we reduced BTU/square foot/heating degree day energy consumption at our schools, from double digits, 12 to 17, to single digits, 5 to 8.

If you what to know more about HVAC system operations; reduce your company's energy cost and carbon foot print; increase comfort and indoor air quality; as well as increase the number of accolades from your supervisor, building users and your company's bean counter, you can do no better than learning from Tom Olson, a knowledgeable and passionate teacher.

Larry Lutz
Director of Facilities and Maintenance
Riverton Community Housing

Preface

When my son was in high school, he said, "With all that you know about the HVAC industry, why don't you write a book?" Well, I didn't have time to write a book. But, later in my career, when the Internet was invented, and I was finally comfortable with using e-mail, I decided that it would be easy to author a book, if I just wrote a page a month. If I then e-mailed it to every client, and prospective client, whose e-mail address I could get, it may help me build a relationship where they would come to know, like and trust me and my company. Five years later, having never missed a monthly installment, I believe my page-at-a-time book had done just that. Did all 1,500 readers become customers? No, but that wasn't the goal.

I am now several years into my retirement from the HVAC service and building automation business that I co-owned. As I continue to do occasional consulting, I realize that there is still a need to share my experiences with everyone that wants to optimize HVAC energy consumption, as well as improve comfort and indoor air quality.

So, for those that want help accomplishing more with less, I've taken my page-at-a-time book; updated the information; added photos from some of my presentations; and made it into a book. I'm convinced that 75% of all HVAC maintenance is simply keeping the HVAC system clean, dry, and lubricated. I believe no one can provide those services more cost-effectively than well-trained in-house personnel. If you need to do more with less, I believe this book can help.

In 1992, I participated in a University of Minnesota sponsored study to determine what percentage of current energy consumption was being wasted in commercial office buildings. They started by asking three different focus groups what they thought that number would be. Of the nearly 50 total participants, only one person said the number was more than 10%. I said the minimum number was, **at least**, 20%! The other participants at my roundtable event were shocked at my number. The engineer that sat next to me told me I was crazy.

Long story short, I was also asked to review the data that was collected in the study that followed. As you might imagine, it came as no surprise to me that the study concluded that virtually every commercial building had an opportunity to reduce their HVAC consumption by 20%, simply by returning the systems to their originally intended operating condition! **Fifteen years later**, in 2007, The Building Owners and Managers Association, BOMA, challenged their members to reduce their energy consumption **30%**, by 2012! After all that time, a large percentage of commercial buildings still have savings of that magnitude, and more. So, no matter what type of facility for which you are responsible, there is a substantial opportunity to reduce the energy consumption, and in all likelihood, increase comfort and indoor air quality.

In 1995, our nation's K-12 public schools admitted to having a $112 billion deferred maintenance opportunity. Much of that deferred maintenance affects energy consumption, comfort, and indoor air quality. In a 2002 United States Department of Energy booklet, it was suggested that America's schools could reduce their energy consumption by an astonishing 25%! (Author's note: My former company was responsible for helping scores of facilities reduce their heating energy consumption by much more than 25%! I believe 25% is conservative.) In real money, that's $1.5 billion per year. In numbers of teachers, that's 30,000 new teachers in classrooms. If I've done the calculation correctly, that's also over 15 billion pounds of carbon dioxide, not put into the atmosphere, every year, just from America's public schools.

So that you'll have confidence in the information provided, some background information is in order. I have a Bachelor of Aeronautical Engineering degree, from the University of Minnesota. My first job out of college, in the aerospace industry, wasn't that for which I was hoping. So, after three years, I completely changed industries. When I first joined the HVAC industry, I was being trained to sell HVAC services, because they were necessary for comfort, indoor air quality and long equipment life. Within one month, we experienced the 1973 oil embargo. All of a sudden, it was all about selling improved energy efficiency. When heating oil was $0.11 per gallon, and natural gas was competitively priced, energy efficiency wasn't a big deal. But overnight, the higher cost of energy was the driving force behind nearly everything most HVAC companies did.

From a temperature control perspective, where I really got a leg up on my competitors, was as a member of the Minnesota Chapter of ASHRAE (The American Society of Heating, Refrigerating and Air-Conditioning Engineers) Energy Committee. In 1975, the committee was charged with reviewing and commenting on a document called ASHRAE 90-75. When the nationwide review and comment period was completed, this document became the Energy Code adopted by all 50 states. I was the only control specialist in the MN group and it really helped me understand that the air-handling unit sequences of operation, at that time, were providing costly and uncomfortable simultaneous cooling and heating.

After the first round of required energy audits for Minnesota's public schools, in 1979, I discovered that only one of these energy audits ever addressed the need for improved temperature control sequences. So, I set out to change that. At my expense, over the course of several weeks, I invited over 250 Minnesota energy auditors to a free class, and box supper, at my office. The person in charge of reviewing the second round of audits said that he could tell every auditor that attended my class.

Early in my HVAC career, I was convinced that if I could just show people what I saw day in and day out, they would better understand that there needed to be a working relationship between HVAC servicemen and in-house maintenance personnel. There is plenty of work to do in all schools and commercial facilities. There was no need for my servicemen, or the in-house personnel, to view each other as competitors. Everyone could accomplish more as partners, than competitors. So, I started carrying a camera with me. It didn't take long before I had a pretty impressive slide show. I've given that presentation twice at the Minnesota Chapter of ASHRAE, once at a Chapters Regional Conference (CRC) for the Central Region of ASHRAE, and scores of school and BOMA training programs. I believe I was the only contractor ever asked to chair the St. Paul, MN BOMA engineers group.

After a number of committee assignments, I was elected to the positions of treasurer, secretary, president-elect and president of the Minnesota Chapter of ASHRAE. I have received the Young Engineer of the Year Award for the Minnesota Chapter of ASHRAE and Rotarian of the Year from my local Rotary Chapter.

At a Wisconsin Association of School Business Officials (WASBO) meeting, Mr. Terry Pease, Director of the Wisconsin Energy Initiative – 2, told the group that my e-mail lessons were the best, ongoing training program that he had ever seen. Thanks, Terry! I hope you now enjoy reading those lessons, all in one place.

Introduction

With over 35 years' experience in the HVAC industry, one of the most universal problems I've found in schools, and commercial facilities, is underfunded maintenance departments. This often leads to the very problems this book is intended to help you solve. In this book, you will be guided to establish proper maintenance funding and energy budgets. It will also help teach your in-house maintenance personnel how to perform as much as 75% of all HVAC maintenance services, saving you thousands of dollars a year.

While this book will give in-house personnel a jump start on being able to perform the majority of all HVAC service needs, there is still 25% for which you will likely need specialized contractor services. I encourage in-house maintenance personnel to form an effective partnership with the contractors that work with HVAC systems day in and day out. That means no secrets! If your contractors are **unwilling** to work with, and share their knowledge with your in-house personnel, keep looking until you find contractors that are. The only reason for not sharing is that the serviceperson doesn't know what he/she is doing well enough, or he/she doesn't know where to get the information, as quickly as he/she should. At the same time, know that being able to perform 75%, or more, of the needed HVAC system maintenance is better than about 90% of all other schools and commercial facilities.

I personally once saw the following sign in a high school shop class:

> **I hear and I forget**
> **I see and I remember**
> **I do and I understand**

Initially, the partnership with your contractors may slow the work they are performing, but once your in-house personnel "understand", the rewards can be limitless! I know no one better capable of providing most of those services, at the lowest possible cost, than the facilities' own well-trained in-house personnel. The initial lessons that I e-mailed were an expansion on a six-step process that, combined with the active partnership of your contractors, can greatly improve the capabilities of your in-house personnel. My former company has demonstrated the effectiveness of the six-step process for decades.

One of the biggest deterrents to accepting my recommendations for energy savings was the disbelief in my savings projections. It's that old saying, "If it sounds too good to be true..." As the following example demonstrates, I believe my projections were more right than wrong.

I was trying to help an eight building school district. At one time, they had heating energy consumption rates as high as 26.0 Btu/square foot/heating degree day. When I first approached them, their average heating energy consumption rate was approximately 13.0 Btu/square foot/heating degree day. Because all the facilities were steam heated, I suggested a modest target heating energy consumption rate of 7.0 Btu/square foot/heating degree day for each facility. "We've already reduced our consumption by 50%. How could you possibly help us anymore?"

Finding additional savings projections of that magnitude hard to believe, they were quite reluctant. So, I asked them to please let me give them a tour of the facility of their choice. With the director of buildings and grounds, his assistant and the district's business manager, we spent a couple of hours touring the boiler room, a couple classrooms and an air-handling unit equipment room. After showing them, first hand, the opportunities to improve the operation of their HVAC system, we started working immediately. I am happy to report that, in time, every building in the district operated at, or below 7.0 Btu/square foot/heating degree day.

The process used to optimize energy consumption, improve comfort and indoor air quality is a simple six-step process:

1. Fix what's broken

It's impossible to provide the desired efficiency, comfort and indoor air quality with broken or damaged equipment. If it's a needed part of the HVAC system, and it's broken, it should be on a priority list for repair or replacement. I'll help you better understand what to look for and why the repairs are important.

2. Clean what's dirty

Dirt and debris are your HVAC system's biggest enemies. It causes premature equipment failure, inefficiencies, and indoor air quality related problems. When in the aerospace industry, I worked in a manufacturing facility for Titan 3-D missiles. You could literally eat off the floor. So, when I joined the HVAC industry, the filth that I found was shocking! How can these systems possibly operate efficiently with all that dirt? Well, they can't. I'll spend a great deal of time talking about getting, and keeping, HVAC systems clean.

3. Change methods of operations made possible, because the system is no longer broken and dirty

Building maintenance staff will be more productive and efficient, when given the knowledge, and the time, to operate a facility free of broken and dirty equipment. Often times, however, they've never had an opportunity to operate a building free of such deficiencies. Methods of how to operate a facility to cover up for such deficiencies frequently just get passed from generation to generation. This book is intended to help break that chain of events.

4. Temperature control system revisions

Energy efficiency, comfort improvement and indoor air quality are not mutually exclusive terms. Existing, antiquated control sequences are often the root cause of preventing success in these areas. It is important to utilize proven, modern control sequences to eliminate simultaneous cooling and heating. From what you learn in this book, you'll be able to recognize improper air-handling unit operation and you'll even be able to share ways to improve them with your temperature control contractor.

5. Install new technology hardware

It is frequently in your best, long-term interest to replace defective equipment, instead of investing in repairs of old, antiquated equipment. In many instances, there are new, unique and often low-cost equipment solutions. I'll share many ideas with you.

6. Implement preventive maintenance routines

Preventative maintenance is an area that should be incorporated throughout the six-step process. A good schedule of preventative maintenance is one of the most important factors in managing time and financial resources. Again, I will help your in-house maintenance personnel become more self-sufficient.

When it came time to start writing my page-at-a-time book, I simply expanded the list of items that came to mind, when reviewing the above outline. The intent of the lessons e-mailed was to have each lesson take only a minute, or two, to read, so many of the lessons took more than one e-mail. Then, with other special notes, these lessons kept us in contact with our clients, and prospective clients, for over 5 years! Now, you have that information as a ready reference. Thanks for allowing me to help.

While every effort is made to provide dependable information, the author, printer and editors cannot be held responsible for any errors or omissions.

Fix what's broken

1.A Temperature Control Air Compressor

My company started in 1978 as a pneumatic temperature control service company. Properly done, pneumatic temperature control system improvements can be nearly as effective as today's electronic building automation systems (BAS). The primary issue is, as today's pneumatic control specialists retire, they are not being replaced. Still, there are a significant percentage of schools, hospitals and commercial office buildings that continue to operate with pneumatic controls. Until such time that these systems can be replaced, the following details will help extend their life.

AIR COMPRESSOR OPERATING INSTRUCTIONS

Critical to the proper operation of a pneumatic temperature control system is a clean, dry, oil-free source of air. The heart of a pneumatic temperature control system is the air compressor. The following are guidelines, which I recommend you follow closely.

Operate air compressor systems on a year-round basis. You do not want warm, moist air entering the airlines when the compressor is off. Zero air pressure also adversely affects control diaphragms and more.

Drain water from the compressor air tank daily, year-round. Make certain that the air is drained from the bottom of the tank, not the midpoint on the end of the tank, through a siphon. I have seen numerous air tanks where the siphon has broken, or rusted off, effectively cutting the size of the air tank by 50%. Have the drain line from the tank, and the refrigerated air dryer, piped to a container. You want to be able to view what's being drained, as well as keep possible oil contamination out of the floor drain and off the floor.

Keep an eye on the liquid in the drain-water container. Is it clear water? That's what it's supposed to look like. If it's white or milky looking, this indicates the presence of oil contamination. Most temperature control air compressors are oil lubricated. It is important to make certain that the air compressor is specifically designed for temperature control use and that you guard against oil contamination. That may cause some to consider the use of oil-less air compressors. I caution against it. As troublesome as oil contamination may be, it pales by comparison to the airborne debris from the Teflon rings of oil-less air compressors.

The optimum design for a temperature control air compressor is 4 minutes on - 8 minutes off. This extended off time allows the compressor cylinder(s) to cool. This total on/off cycle time adds up to 12 minutes, or 5 on/off cycles per hour.

Air compressor systems, with more than 5 on/off cycles per hour, have a need for further investigation. Either the tank is too small; you paid too much for an oversized compressor; both of the above; the on/off pressure differential is too small; or the system has lots of air leaks. Again, the compressor cylinder(s) will not cool if the system has too frequent on/off cycles. This heats the lubricating oil to the point of vaporizing. As that oil is carried throughout the system, it condenses, plugging restrictors and damaging control devices. Keeping the oil from entering the airlines is much less expensive than removing it after the fact.

Another major concern is the run time ratio. Air compressor systems, whose on time exceeds the off time, either have a compressor that is worn out, too small or the system has too many leaks. Again, oil contamination is a probable result.

Between the air compressor and the refrigerated air dryer, there should be no filters or pressure reducing valves. Moisture is certain to condense at these points, shortening the life of these devices.

A properly sized, good quality, refrigerated air dryer is a must. When at the compressor, draining the tank on a daily basis, grab a hold of the outlet of the refrigerated air dryer. It should feel either cold or hot to the touch. Let me explain. The refrigerated air dyers of years ago used to have a separate condensing coil for the refrigerant. That means, that the heat of compression is put back into the space, making the airline leaving the refrigerated air dryer feel cold, when operating properly. Be certain to look at both sides of the condenser coils to see if it needs cleaning. Vacuum it clean.

Today, many refrigerated air dryers use the cold air from the air dryer to condense the refrigerant. For those dryers, the outlet airline should feel hot to the touch. So, if the outlet airline is neither cold, nor hot, seek immediate assistance. You simply can't afford to have moisture in your pneumatic temperature control systems.

The minimum air compressor tank pressure should be 60#. This allows the air to be in the refrigerated air dryer long enough to cool to the condensing temperature and wring out the moisture.

Clearance between the wall, and the compressor flywheel, should be a minimum of 14". This allows the flywheel to operate as a fan and help cool the cylinder(s). Be certain that the flywheel rotation blows the air across the compressor. It cools the compressor better than trying to draw the air across the compressor. If the rotation is wrong, it has something to do with switching the power wires. It depends on if it's single phase or 3-phase electricity, so be certain to consult with someone more knowledgeable than I.

Use two air intake filter elements for the compressor. Alternate them twice a year - minimum. More frequently, if the area is dirty. Use good quality, washable type filters. Use a mild liquid soap solution, rinse thoroughly and put the freshly washed filter in a clean, dry location between changes.

Air compressor oil should be changed only once a year. Air compressors do not develop the same levels of heat that a combustion engine does, so the oil does not have to be changed as frequently. Remove the access plate and clean the sludge out of the bottom of oil sump, at every oil change. For pressure lubricated air compressors, which I highly recommend, clean the oil pump pickup screen at the same time. Be certain to have a new access plate gasket on hand. Replace the access plate gasket, in case it was damaged in removal.

Use a good quality, heavy-duty air compressor grade oil only. Do not use any type of automotive grade oil. Not even, non-detergent automotive oil. Automotive oil is designed to go everywhere. That's good for cars, but it is not what is needed for temperature control air compressors.

When a temperature control air compressor fails, replace it immediately, using a slow RPM (<400 RPM), pressure lubricated compressor, specifically designed for temperature control use. Quincy is the only such major manufacturer of temperature control air compressors, with side-seal rings, of which I am aware. The ends of the rings are made at opposing angles, so that they overlap. My experience is that a few extra dollars, in a properly sized air compressor of this style, is well worth the investment.

Many of today's refrigerated air dryers have a dirt filter, as an integral part of the water dump mechanism. This operates as a pre-filter, extending the life of the oil filters that follow. To determine whether or not this filter is dirty, there should be air gauges on the entering and leaving sides of the air dryer. Most of the operating manuals indicate that, when the pressure drop across the gauges reaches 10#, the filter should be cleaned. My experience indicates that cleaning is a waste of time. The cost of a replacement dirt filter is minimal. Go ahead and replace the dirt filter, when the pressure drop indicates that it is time.

After the dirt filter, there should be a good quality, coalescent oil filter. Again, this filter requires air gauges on each side to determine the pressure drop. The air gauge after the internal dirt filter, mentioned above, acts as the air gauge on the entering side of the coalescent filter. The coalescent oil filter is also replaced when the pressure drop between the gauges reaches 10#. As a control contractor, we had a great relationship with the local Norgren Company representative that provided us with all of our oil filters. They were also the Quincy air compressor dealer.

The coalescent filter should be followed by a superior quality, activated charcoal, oil vapor adsorbing filter. This is your last line of protection. Don't try to save a few pennies here. The frequency, with which this filter is changed, is determined by the filter quality and the system air consumption. Your installing contractor, or Norgren filter supplier, should help you determine the frequency of changing this filter. As a general rule, even if the air consumption is low, this filter should be changed, at least, annually. I suggest it be changed when changing the oil on the compressor.

Remember, all of the above filters are to be at high pressure (60# minimum) and after the refrigerated air dryer.

Feel the outlet line of the refrigerated air dryer every day. It should feel cold, like a glass of ice water, or hot, like a cup of coffee. In the summer, you may even see a cold line sweat, like a glass of ice water. If it's doing that, you don't even have to feel it. If it is neither cold nor hot to the touch, immediate adjustment, repair, or replacement of the refrigerated air dryer is recommended. The key word being immediate. Moisture in pneumatic air lines can cause lots of unpleasant and costly damage.

After all the filters, you will find one, or more, pressure-reducing valves (PRV's). Some systems have one PRV for the day setting of the thermostats and another for the night setting. Any adjustments to the pressure settings must be done when there is air flow through that PRV. That is, only adjust the day PRV, when in day operation. After switching from day operation to night operation, wait a few minutes to let the system settle out, before making any adjustments to the night PRV.

Following every PRV, there should be a 30# safety pop-off valve. This is needed, in the event of a failure of the PRV. At least once a year, typically when changing the compressor oil, these pop-off valves should be exercised. Wrap the pop-off valve with a tan paper towel. Pull the ring, on top of the safety valve, until it blows air at full force. If there is moisture or oil at that point, the paper towel will help keep it from blowing all over everything, including you. It will also help in analyzing what's going on.

If the towel appears wet, set it aside for a while. Later, if the towel has dried to its original color, the dark stain was just water. Get the air dryer checked immediately. If the dark stain is still there, it's oil contamination!

Double-check all the filters, replacing as required. If the problem isn't obvious, recheck the compressor run time; the distance between the wall and the flywheel; make certain that the flywheel rotation is correct; and so on. If the air compressor is not properly cooled, it will pump oil.

Removal of oil contamination is extremely expensive. Not removing oil contamination can be even more expensive. Oil contamination can cause poor operation of controls, resulting in lots of comfort and indoor air quality complaints, inefficient operation, constant recalibration and potentially costly replacement of damaged controls. The last time my company was involved in removal of oil contamination, however, it required the use of Freon TF, a chlorofluorocarbon. You'll want to contact your local pneumatic temperature control contractor for help with this. There may be something new in oil removal. If your contractor doesn't know how to remove oil contamination, call another contractor. You should eventually find one that knows what they're doing. That's the one I'd recommend to do the control revisions too. If needed, encourage that contractor to buy a copy of this book.

Instead of going to the expense of removing oil contamination, repairing or replacing defective controls, and upgrading the sequence of operation of pneumatic temperature controls, this is where you need to start looking at the big picture. If the pneumatic temperature controls are broken, dirty, contaminated with oil, and providing poor sequences of operation, you may be better off taking the money needed to refurbish the pneumatic controls and investing in a modern building automation system, BAS.

When a replacement air compressor is needed, do not buy a new compressor that is **not** specifically designed for temperature control use. Again, I recommend a slow speed, pressure lubricated, **Quincy** air compressor, specifically designed for pneumatic temperature controls. Properly sized, they can last for decades. Anything less than a properly sized air compressor and storage tank system, specifically designed for temperature control use, will cost you more in the long run. Replacement air compressor and tank assemblies should be sized for one-third run times and five on/off operations per hour. That is, 4 minutes on, and 8 minutes off, is an ideally sized air compressor and tank assembly. This assures that you spend enough, without spending too much.

Keep good records on the compressor on/off run times, down to the second. It will help to determine the cause of current and future air compressor problems. If the on time goes up, but the off time remains the same, your compressor is probably getting weak. If both the on and off times go up, you may have new air leaks in the system. See Chapter 3.I on use of ultrasonic fault detectors for finding air leaks.

Now, here is the information you need to properly size a new compressor:

To determine the air usage of an existing pneumatic temperature control system, first determine the size of the air tank.

> 14" X 33" = 20 gallons
> 16" X 38" = 30 gallons
> 20" X 48" = 60 gallons
> 20" X 63" = 80 gallons
> 24" X 69" = 120 gallons

Next, measure the time it takes for any of the following pressure drops. Use the largest pressure drop possible to get greater accuracy, even if you have to turn the compressor off to get it.

Cubic Feet of air @ various pressure drops*

Gallons	25#	20#	15#
20	4.55	3.64	2.73
30	6.83	5.46	4.1
60	13.6	10.88	8.16
80	18.1	14.48	10.86

Now, use the following formula to calculate system usage:

Pressure drop volume/pressure drop time (as a decimal) = Cubic Feet per Minute (CFM)

Example:

30 gallon tank takes 2 minutes, 30 seconds to drop 20#
30 seconds/60 seconds = 0.5 minutes
5.46 cubic feet/2.5 minutes = 2.184 CFM air usage

Now, because the compressor will be running, while still using air, multiply the air usage times 3.5 to get the compressor size needed for a one-third run time.

2.184 CFM X 3.5 = 7.644 CFM
Okay, that's the size of the compressor you need for a one-third compressor run time.

Now, just because the compressor will run just one third of the time, it doesn't mean that it will give you the desired five on/off cycles per hour. So, now take the air consumption of 2.184 CFM X 60 minutes = 131.04 CF per hour.

Next divide 131.04 CF by 5 cycles = 26.2 CF per cycle.

Looking at the above tank chart, you see that, with a 25# pressure drop, if you add the volume of an 80 gallon tank, 18.1 CF, and a 30 gallon tank, 6.83 CF, you'd be just a little short with 24.93 CF. Fortunately, the next size tank is 120 gallons, (120/80 x 18.1 CF = 27.15 CF), so that's the tank size you'd want with a 7.644 CFM, or slightly larger compressor, for this example. If space is a consideration, two 60 gallon tanks are the same as one 120 gallon tank.

Finally, ask your control contractor, or air compressor provider, for an air compressor that delivers that much air on that sized tank(s). Your contractor, and maybe even your compressor supplier, will want to know how you calculated it. Thanks for recommending that they buy a copy of this book. There's a lot more in the book for them too!

Quincy Compressor
3501 Wismann Lane
Quincy, IL 62305-3116
Phone: 217.222.7700 (8-5 M-F CST)
www.quincycompressor.com

Norgren Air Filters

External dirt filter – F74G, Sintered polypropylene dirt filter element

Coalescent oil filter – F74C, Synthetic fiber and polyurethane foam element

Oil vapor filter – F74V, Activated carbon and aluminum element

Norgren, Inc.
5400 S Delaware
Littleton, Colorado, 80120
www.norgren.com
Harry Milton
HMilton@usa.norgren.com

Fix what's broken

1.B - Steam Traps

After assuring a clean, dry, oil-free source of air for pneumatic temperature controls, the items most frequently found in need of repair are steam traps.

So, how do you test steam traps? I'll go into greater detain in Chapter 3.I, Ultrasonic fault detectors, but sometimes, knowing that you have steam trap issues is as easy as just paying attention to what's going on at the condensate return tank, in the boiler room.

Figure 1.B.1 Steam from defective steam traps blowing out of condensate tank

The next photo is from a public facility that found an easier answer to the work that would be required to repair or replace all the defective steam traps. When they found steam blowing from the condensate tank, as above, they simply ran cold water from a garden hose over the tank - 24/7! The tank is covered with a thick layer of rust on the tank, so they'd been doing this for some time. What a waste!

Figure 1.B.2 Cold water running on rusty condensate tank to keep steam from filling the boiler room

If you're going to do steam trap testing, and repair or replacement of defective traps, in the summer months, you'll need to have steam available. Some schools do not want to disturb classrooms during the school year, so they wait until school is out for testing steam traps. Except for some wasted energy, and a little extra unneeded heat in the spaces, there is no problem in doing this. If that's what you plan to do, just be certain to communicate that with your building engineers, so the boilers are not torn down for the summer, before you start.

The average life expectancy of most thermostatic steam trap elements is just three heating seasons. This applies to all thermostatic elements, whether in thermostatic steam traps, or float and thermostatic (F&T) steam traps. The life expectancy of most F&T float mechanisms is about six heating seasons.

The economic justification for replacing defective thermostatic steam traps is phenomenal. In 1995, the Energy Division of Minnesota's Department of Public Service did an analysis of the energy savings for steam trap repair or replacement. Thermostatic steam traps are usually found at the end of small heating coils, including radiation. A control valve generally modulates the flow of steam into the coil, so that the steam is often fully condensed, even before it gets to the steam trap. This analysis assumed that the control valve would be closed during night setback times, whenever the space temperature was above the setback temperature. The analysis further assumed that 50% of all steam that blew through defective steam traps still added useful heat to the building. I disagree, but it makes the analysis all the more conservative. Updating that conservative analysis to $0.70 per Therm (100,000 Btu), a natural gas user would save approximately $210 per steam trap, for an average Minnesota heating season. In virtually every case, including installation, that's less than a one year return on the investment (ROI).

F&T steam traps are found on steam line headers in the boiler room; in end-of-main steam distribution lines; and on larger heating coils. The savings from repairing or replacing a float and thermostatic steam trap, on a coil, with a control valve on the entering side of the coil, is the same as a thermostatic steam trap - about $210 per year, per steam trap. Most will still have a one year return on the investment. Even larger F&T traps pay back rapidly.

In boiler room header and main steam line applications, these F&T steam traps do not have a control valve ahead of them to limit the amount of steam that is wasted, when the trap is defective. According to the Minnesota analysis, these worn out steam traps cost the user approximately $1,050 per steam trap, per average heating season!! Again, this is based on natural gas at $0.70 per Therm. This study also limited the number of heating season hours to only the coldest 4,900 heating hours, instead of the usual 6,000 hours spent below 60F, in Minnesota. So, these numbers are quite conservative. The cost of replacing an end-of-main steam trap, with a long-life steam trap, is generally less than $500, including labor!

With all that short-life steam trap information, it's not all bad news. There have been substantial developments in steam trap technology.

Do you remember the last time that you had to replace the thermostat in your automobile radiator? I think it was 1967 for me. In the late 1950's and early 1960's, changing automobile radiator thermostats was a semi-annual ritual. Learning how to perform that task was a like a rite of passage for every young man. The radiator thermostats, at that time, were very similar to the bellows style element found in most thermostatic steam traps made yet today. After a few thousand cycles, they simply fail from metal fatigue. And, it's understandable, once you realize that it's like taking a piece of tin and bending it back and forth until it cracks.

Today's automobile radiator thermostat has a paraffin filled sensing element that pushes a gateway open to allow flow past the sensing element. That same style element is also now available in thermostatic steam traps by a company called TLV. A nationwide steam trap survey in 1974, indicated that 0% of thermostatic steam traps were functional at the end of 5 years. I've of the paraffin style steam traps still going strong after 10 years.

service company would install these TLV thermostatic traps, we guaranteed them against the loss of live steam for 3 years. It never cost us a dime!

F&T steam traps have also seen significant technological improvements. In most, older F&T traps, the ball in the float mechanism is comprised of two hemispheres, simply soldered together. To the ball, an arm and pivoted extension are attached. The pivot often gets sloppy. The plug at the end of the extension gets worn. The seat, to which the plug is supposed to seal, also gets worn.

The biggest improvement in F&T steam traps is one that is, again, manufactured by TLV. The float mechanism is a seamless, perfectly spherical, free floating, stainless steel ball. I wish I had been able to figure out how to make a ball like that. The float has no arms or levers to wear out. The ball self-hinges on a hardened stainless steel seat, which is below the waterline of the condensate, rising and falling as the flow of condensate varies. With no new condensate entering the trap, the ball perfectly seals against the seat. The thermostatic element is a long-lasting bimetal.

I've seen this free-floating ball style steam trap provide trouble-free operation, for more than ten years, in year-round operations, at 250 # steam pressure. I believe they will last much longer in low-pressure applications. When my company installed them in HVAC systems, we'd guarantee them against the loss of live steam for ten years! That's unheard of in the HVAC industry. Again, they never cost us a dime. I'll provide TLV contact information at the end of this chapter. Because of the life expectancy of all TLV steam traps, I recommend the replacement of all defective steam traps. Repair just isn't worth the effort.

Two other areas where I see lots of energy waste, and comfort problems, are on entrance fan coils and unit heaters. This equipment is often installed without the benefit of a control valve. On a call for heat, a thermostat cycles the fan on, if there is steam in the heating coil. The installation of a simple, self-contained radiator valve is the usual low cost solution to the overheating. Remember too, the steam traps on these uncontrolled units can waste just as much steam as a defective end-of-main trap - about $1,050 a year!

While talking about long-life steam traps, I also have to state that they require a proper water treatment program. I once had a client from the most remote part of Minnesota. He said he had a steam trap location that he couldn't get the trap to last more than a few months. So, I sized up a new TLV steam trap and sent it to him. A year later, he called to say that our #%!*&^ trap had also gone bad. I asked him to send it back and we'd examine it. Upon inspection, we found that the hardened stainless steel seat was totally gone! Yes, GONE!!

I called him back to ask about his water treatment program. He said, "Never mind my water treatment program. The problem is your #%!*&^ TLV trap." Knowing lots of water treatment people that could possibly help him, despite his remote location, I asked, "If I can get someone to come all that distance (300 miles), would you let them take a water sample?" Reluctantly, he finally agreed.

A couple weeks later, the water treatment specialist called to tell me that the school's boiler water tested at 0.9 pH!! As I'm certain you know, 7.0 is neutral. Not only was his boiler water so acidic that it was eating away the TLV steam trap, it was eating away all metals with which it came in contact! That's boilers and piping! Build a good relationship with your water treatment specialist. They can save your bacon!

Minnesota natural gas utilities provide rebates for the repair or replacement of defective steam traps and other items. We encourage you to take advantage of rebates, as long as they last. Check with your natural gas provider to see if rebates are available in your area. With or without a rebate, however, keeping your steam traps in proper repair is a great investment.

Here is another actual story about steam trap replacement. I had convinced the director of buildings and grounds to let us test the steam traps in one of his elementary buildings. The building was over 400 feet long and the boilers were at the north end. We quickly determined that testing the traps was a waste of money. After several hours of testing, we had yet to find a single working steam trap. So, the director agreed to have us start replacing all the traps, immediately.

The facility engineer wasn't very happy that his boss was allowing us to do all that work. His theory was that he needed 8 # steam pressure, at the boilers, in order to have any heat at the far end of the building. And, up to that point, he was probably right. He didn't see how our work would allow him to lower that pressure to our recommended 5 # maximum pressure. To prove his point, he installed a steam pressure gauge at the far end of the building, opposite the boiler room. When we were finished, he checked the pressure at both ends of the building. Both showed 8 #. So, he lowered the pressure to 5 # and still saw no difference. He kept lowering the pressure until, finally, when lowered to just 2 #, he said he saw a ¼ # drop in pressure from the boiler room to the far end.
The following is a photo of float elements from float and thermostatic steam traps in this facility:

Figure 1.B.3 Actual float mechanisms from an elementary school

The float at the back of the picture developed a hole. The maintenance personnel brazed the hole, but not very successfully. The patch fell off and the float filled with water and sunk.

The float on the right never did have the water removed. With all of the brazing, it couldn't float.

Finally, the float in two pieces was so severely impacted by water hammer that it was literally blown apart. Obviously, it's days of floating were long over.

When we first started this project, the building engineer saw us as a competitor. Decades later, after having a close working relationship, he finally recalled this event and admitted to having lied to me. He never did see any drop in the pressure, 400 feet from the boiler room. Not that we asked him to lower it less than 5#, but he said he just wasn't comfortable trying to drop the pressure any further than 2#. That project paid for itself in less than one year and led to similar projects throughout the district.

Unless replaced with long-lasting, TLV steam traps, every steam trap, in every facility, should be tested annually. Every one of my company's service personnel had an ultrasonic fault detector in their truck. Ultrasonic fault detectors are one of the most beneficial tools that any maintenance department could ever use. Annual testing is especially critical for the end-of-main traps, where the savings could exceed $1,050 per year! I recommend steam trap testing early in the spring. This is when most maintenance departments, and service contractors, have a little breather between the rigors of the heating season and the hectic pace of summer projects. With the list of defective traps in hand, repair parts, or new, long-life steam traps, can be ordered and the work can be scheduled for completion before the heating season starts all over again.

The exception to testing steam traps before replacing them is when it has been more than 5 years since any testing and repairs have been done. Under those circumstances, just replace them all. You can't justify investing in finding an occasional trap that may still be functioning.

Now, after having gone through all of that discussion on steam traps, I'm going to suggest that, if your facilities have a life expectancy of 15 years, or more, you should investigate the possibility of converting the steam heating system to hot water. Done without having to change out all the piping, and maybe even some control valves and heating coils, it's probably not as costly as you might imagine. The energy savings of 30% to 50%, plus the reduced maintenance costs of up to 72%, according to one study, certainly make conversion to hot water heat something to look into. I'll have much more on this subject in Chapter 5.B, Boilers, including modern hot water reset and Chapter 6.D, Benchmarking.

TLV contact information:

TLV Corporation
13901 South Lakes Drive
Charlotte, NC 28271-6790
Phone: 1-704-597-9070
Fax: 1-704-583-1610
www.tlv.com

Fix what's broken

1.C - Air-Handling Units

Outside air intakes

Figure 1.C.1 Partially blocked outside air intakes

This is pretty classic. On the entering side of every outside air, OA, intake is a rain louver. Like its name implies, it is designed to keep the rain and snow from entering the air stream. OA intakes are sized so that the entering air does not exceed 500 feet per minute, fpm, velocity. Anything faster than that and the outside air will drag in snow and rain. As an example, a 25,000 cubic feet per, cfm, air-handling unit that, at times, brings in 100% OA. 25,000 cfm, divided by 500 fpm equals 50 square feet (sf). That's the size unrestricted OA intake this unit needs, to keep from sucking in rain or snow. Now, as illustrated in the above photo, imagine plugging 80% of the OA intake. The air-handling unit still tries to provide the same 25,000 cfm. Divide 25,000 cfm by 10 sf (20% of 50 sf) and the air velocity increases to 2,500 fpm. Wet air-handling unit interiors are a certainty. If you've ever been a part of such action, please don't ever do it again.

Figure 1.C.2 The light colored rectangles are outside air intakes for unit vents that are boarded closed

In the early days of skyrocketing heating energy costs, ASHRAE reduced the ventilation requirements from 15 cfm to 5 cfm. (Note: They are back to 15 cfm.) At the same time, the above school was experiencing numerous frozen hot water heating coils. So, they just decided to totally block the OA intakes to all their unit ventilators. It's important to understand how dampers work to see just how this impacts the unit's operation.

When the control signal to OA and return air, RA, dampers is zero, or the signal fails, the dampers are supposed to go to their normal condition. For OA dampers, the normal condition is closed, to aid in freeze protection. The RA damper is linked to the OA damper and operates in the opposite direction of the OA damper. It's also important to remember that school classrooms have a tremendous amount of heat gain. So, unit vents are trying to cool the classroom during most occupied times of the day. That means that the thermostat is trying to bring in OA for cooling. Because, in this case, the OA intake is fully blocked, it is unable to do any cooling. In addition, when the OA damper is fully open, the RA damper is fully closed. That means that the fan is simply freewheeling, moving little to no air at all. Of course, this is an inappropriate operation.

Outside air dampers

When discussing damper problems, it is almost always OA dampers that we're worrying about. The biggest problem, of course, is frozen heating coils. Frozen heating coils require two things – standing water and cold air.

My first rule to keeping heating coils from freezing is to keep unnecessary OA from entering the coil area of the air-handling unit. No, not like in the above two photos. And, it's not even by using top quality OA dampers. In new construction, by all means, specify the best outside air dampers possible. For existing dampers, it's important to take the time to lubricate points of friction and adjust all dampers, for positive closure. But, in practice, I believe it's most important to make certain that all exhaust fans are properly turned off, when not needed.

Air takes the path of least resistance. Certainly with a 40 mph wind blowing at a leaky damper, the danger of freezing coils is ever present. But, exhaust fans sucking air out of the building cause most freeze-ups. The exhausted air has to come from some place and that's the path of least resistance – usually the outside dampers. So, if frozen coils are a problem, the first thing to do is to make certain that all exhaust fans are automatically turned off. Do not rely on humans!

I recommend that all exhaust fans, with daily and regular operations, be turned off automatically, either by a time clock or a building automation system, BAS. For exhaust fan operations that are intermittent and irregular, simple spring wound timers can often be the best answer. Spring wound timers come in a variety of time periods. If 30 minutes is the longest time you need an exhaust fan, select a 0-30 minute timer. If you need 60 minutes, select a 0-60 minute time and so on.

Okay, how about the second ingredient needed for frozen coils – stagnant, or even slow moving water. Let's look at hot water heating systems first. I belong to the theory that "Babbling Brooks don't freeze!" You've certainly seen streams at -20°F that are flowing freely, without freezing. Hot water heating systems should be designed so that there is a continuous flow of water anywhere the coils are subject to freezing conditions. "But, doesn't that waste a lot of energy?" Yes, but only if done incorrectly.

In Chapter 5.C, I'll talk about the economy of improved hot water reset control. It suggests the need to provide heating water just hot enough for the three separate times of occupied, unoccupied and morning boost. Eliminating frozen coils is another reason why I suggest this 3-temperature method of controlling heating water. It's not the temperature of the water that allows it to freeze. It's how fast it's moving that allows it to freeze. Remember, "Babbling Brooks don't freeze!" The solution is as simple as significantly reducing the water temperature at night, drive the heating valves open and keep the pumps on below 35°F.

How about steam coils? How do you keep them from freezing? First, make certain that the control valves are capable of tight closure against the flow of steam. As will be discussed in a Chapter 1.D, steam is a very deteriorating heating medium; control valves are often oversized; and steam systems are operated at pressures higher than the design intended. In addition, building insulation values are often improved, as roofs, and other areas, are repaired, replaced or remodeled. These all contribute to control valves that operate in a mostly closed condition and quickly can't close off to the flow of steam. Ultrasonic testing now makes testing control valves a relatively simple task.

On the other end of the coil, make certain that the steam trap is capable of discharging any condensate that gets that far. Between the control valve and the steam trap, there are vacuum breaker air vents to release the condensate. If they are not operating properly, it's like holding your finger on the end of a drinking straw and taking the straw out of the liquid. We all know what happens when we take our finger off the end of the straw. Yep, the straw is empty. It's the same with condensate in a heating coil.

Again, make certain that all exhaust fans are turned off at night and that the dampers operate properly. Check to see if the existing damper operator has enough power. Then, if all else fails, consider replacing the damper.

Best control story ever

Now I'm going to tell you my very most favorite story about how **not** to operate an air-handling system, complete with a priceless picture.

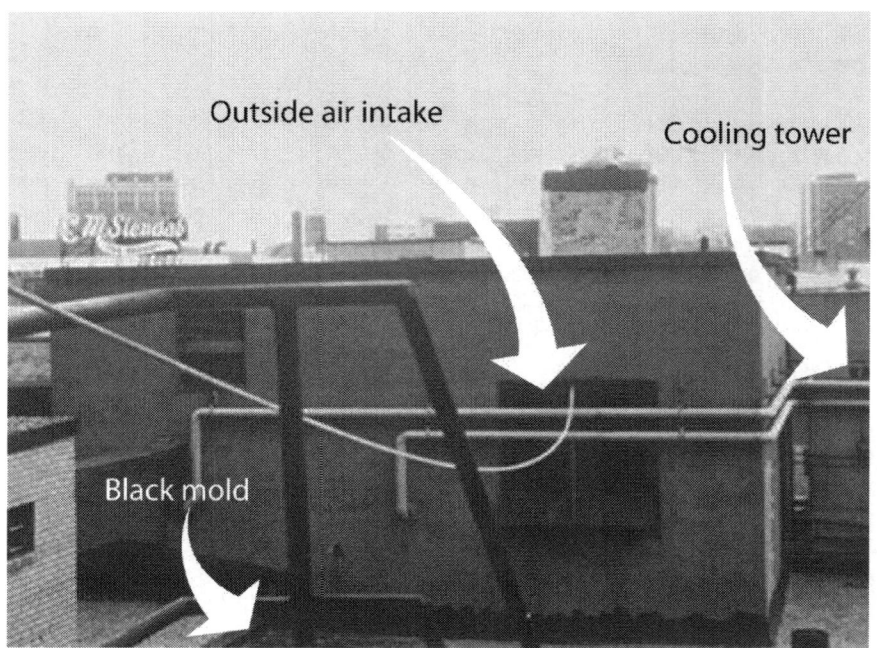

Figure 1.C.3 Best control story ever

We're looking out the window of the maintenance staff office in a medical office building. In transferring this photo, it shrunk just enough to lose some of the detail, but I think you'll understand everything.

To get to the equipment room and cooling tower, you crawl out the window, go down the metal staircase and cross the courtyard. In the building, behind where the pipes go into the building, is a chiller. The black square on the building is the OA intake of the air-handling unit. Off to the right is a cooling tower.

With the building having a short life expectancy, when the heating coil froze and burst, the building owner told the maintenance engineer that he damn well wasn't going to have the controls redone to keep the coil from freezing again. But, he also told the maintenance engineer that he would be damned if it ever froze up again. So, on the 10th floor of a downtown office building, the maintenance engineer installed a ¼" thick steel plate to cover approximately 75% of the OA intake. To that, he hinged another ¼" thick steel plate. To the bottom of the hinged plate, he tied a rope, which he ran up through a pulley and across the courtyard to the office window. When he wanted to bring in OA, he pulled on the rope and tied it up alongside the window. To close the hinged steel plate, he simply untied the rope and gravity did the rest. But, that's just the beginning of this story.

In the front of the equipment room, there was a low spot on the roof adjacent to the roof drain. That meant that not all the water drained off the roof. As with most standing water, it eventually turned moldy. That's the black you can see just below the bottom staircase railing. When the mold dried, the mold spores would spew out and get drawn into the OA intake. Remember, this was a medical office building. Nope, we're still not done.

Remember the cooling tower? It's on the downwind side of the equipment room and the outlet from the top of the tower is lower than the roof. When this picture was taken, the 1976 case of Legionella Disease that killed 34 people in Philadelphia, was still fresh on everyone's mind. That bacterium came from a cooling tower.

Now, imagine the prevailing wind coming off the roof of the equipment room, passing over the outlet of the cooling tower. Then having that air swirl around the building, picking up dried mold spores on its way to the OA intake of the air-handling unit serving the medical office building! I'm glad I never had an appointment there.

Condensate control

More and more today, we see air-conditioning systems having problems with the removal of condensate. Condensate is the water that is removed from the warm, moist air moving across the cold air-conditioning coil. Condensate remaining in the drain pan for more than 24 hours, leads to indoor air quality problems. Provisions must be made to collect and remove this water from air-handling units. Sounds simple enough, but all too often, problems prevent this simple scenario from occurring.

If condensate does not properly drain from an air-handling unit, the pan may overflow, causing water damage to the air-handling unit and everything downstream - the building's structure, finishes and contents. Wet interior air-handling unit surfaces, and ductwork, are breeding grounds for mold and mildew. The cool, damp, dark and dirty surfaces are perfect for growing these health-threatening fungi.

Even when dry, these invaders give off spores and emit volatile organic compounds (VOC's) that cause health problems to building inhabitants, and in some cases, may even cause death. This can be the source of many indoor air quality (IAQ) complaints and liability claims. It is an easy problem to acquire and, up until recently, a difficult one to eliminate. Sure, the moisture dries over the winter, but those little buggers come right back to life, as soon as they get moist again. Worse yet, they continue to emit their damaging garbage into the air, all winter long, even though they are no longer alive!

Controlling condensate from air-conditioning systems is vital to a healthy and well-functioning building. Start by inspecting the condensate pan areas of your air-conditioning systems on the first warm, humid day. Document what's happening. Also check air-handling units ten minutes after they are turned off for the night. There should be no standing water in the condensate drain pan.

There seems to be no end to why condensate remains in some pans, but here are a few:

- To add cooling Btu's to the size of marginal capacity units, some manufacturers actually, intentionally, tilt the drain pan away from the drain to gain some evaporative cooling effect from the standing water.

- Some drain pans are simply too big. Condensate pools in remote areas of the pan, allowing it to stagnate.

- Most air-conditioning coils are on the suction side of the supply fan. The negative pressure on the condensate drain actually prevents the condensate from draining from the pan. In fact, the suction is often so great that the standing condensate geysers into the air stream and is carried well beyond the drain pan.

- Condensate drains dry during the off season, voiding any sealing effect they may have had.

My company's service manager received the following inquiry regarding the use of condensate pan pellets to kill bacteria. "Do they serve any purpose? Some pans always have a little water in them, are the tablets useful then?"

Believe it or not, the problem of dealing with air-conditioning condensate has gotten to be so bad there is actually someone that is the world's leading authority on the getting rid of water in condensate pans. I had a working relationship with this man, Warren Trent, so I dropped him an e-mail note, asking those same questions. This was his response:

"My opinion is that pan pellets are of little or no value. I have seen them in many systems, but have seen no favorable results. Among my contacts in the field, I know of no one, except pellet suppliers, that recommend their use. At best they appear to be an ineffective band-aide."

Given that, there are still people that swear by them. In any case, some of the most effective ways to prevent contaminated condensate pans are as follows:

1. Clean drain pan and treat with an EPA approved biocide.

2. Clean drain lines and treat with an EPA approved biocide.

3. Utilize a narrow drain pan, so that as condensate is formed, it flows continuously over the bottom surfaces of the pan. That is, there are no corners in which water may stagnate. In this type of pan, when condensate flow stops, the residual evaporates before it can stagnate and support biological growth. Large pans may be suitable to be converted to a narrow pan.

4. Slope the drain pan toward the drain, when possible. In some cases, it may be easier to tilt the entire HVAC unit, instead of just the pan.

5. Make certain the condensate drain pipe is properly pitched at a rate of no less than 1/8-inch per foot. Over the years, dirt, algae, etc. will settle out in the drain pipe, slowing the flow and eventually plugging the pipe.

6. Implement a routine preventive maintenance program, which includes periodic inspection of the drain pans and drain lines in action.

The discussion of condensate drains would not be complete without mentioning a product my company used on numerous occasions. Cooling coils can either be on the suction side of the supply fan, or on the discharge side of the fan. Unfortunately, most air-conditioning systems have suction side cooling coils, where most problems occur. The product to solve suction side condensate problems is called the CostGard Condensate Drain Seal. The manufacturer's website has some great information, including a six-minute video. Please take a few minutes to watch it. See contact info at the end of this chapter.

Vibration isolators

These are the canvas, or rubberized canvas connections between the air-handling system and the ductwork. These have a limited life expectancy. They should be inspected annually for the need to be replaced. Annually. tens of thousands of Btu's of cooling and heating can be lost out of these deteriorated connections. Remember, small holes sink big ships!

Duct connections

The same goes for leaking duct connections and ill-fitting doors. Most mechanical rooms do not need heating, cooling or ventilation, because of all the air blowing out of these units. Should you have rooftop air-handling systems, now you're not even keeping this wasted energy in the building. Duct tape is probably just another ineffective band-aide. Talk to you ductwork company about the proper duct sealant to use.

Belts and sheaves

Before leaving the air-handling unit, check the fan belt for its condition and the alignment between the motor and fan sheaves. Also check the position of the belt in the sheave, as well as the condition of the sheave.

Bearings

Just a note here to go to Chapter 3.I for a great story on using an ultrasonic fault detector to keep your bearings in good shape.

Contact information for CostGard Condensate Drain Seal is:

Paul and Warren Trent
Trent Technologies
15939 FM 2493
Tyler, TX 75703
Phone: (903) 509-4843
www.trenttech.com

Fix what's broken

1.D - Control Valves

A control valve is the device ahead of a heat transfer coil, which modulates the flow of steam, hot water or chilled water, to add heating or cooling, as the space requires.

In general, 3-way valves, for control of chilled water, last the longest before needing service. The temperature of the fluid they control is modest and it's not trying to completely close to the flow of the chilled water. If the water doesn't go one way, it goes the other. The most frequent problem is the valve stem packing. It starts to leak. Most manufacturers have packing kits readily available. The packing usually gets damaged from debris collecting on the valve stem. These valves are usually quite large and expensive. Occasional replacement of the packing is a worthwhile, low cost maintenance item. If the packing leaks, it usually causes damage to the pipe insulation and everything else below it.

On some valves, the manufacturer will have an inner valve kit available. It usually includes a new seat and the rest of the valve assembly simply screws out of the valve body, allowing the new inner valve, complete with new packing, to be installed. With proper close-off valves, the entire process takes just minutes.

If installing just a new valve stem packing, be certain to clean the stem with a fine sand cloth. Always clean the stem with a motion up and down the stem. That is, clean it in the same direction of the travel as the stem, never cross-wise. With an occasional new valve stem packing, these valves should practically last the life of the building.

The control valve application that needs the next most frequent service is the 2-way hot water valve. These valves can be as small as ½", or as large as 6". Again, the valve stem packings are often the first visible signs of deterioration. You may also notice overheating, as an early sign of a problem.

During a good portion of the time that I owned the HVAC service company, I was also a manufacturer's representative for Tour and Andersson, a Swedish manufacturer of temperature control equipment. As such, I was heavily involved in the first two hot water district heating systems in the United States. The first was in Willmar, Minnesota. The second was in St. Paul, Minnesota. My involvement in these projects Sweden does not have any domestic source of heating fuel. So, to the fullest extent possible, they use their purchased fuel to produce every ounce of energy possible. They first burn the fuel to make high pressure steam, which drive turbines for the generation of electricity. Once through the turbines, the lower pressure steam is further reduced to condensate, before being reheated back into high pressure steam for the turbines. In the United States, that disposed of heat energy is often times put into lakes and rivers, or is dissipated to the atmosphere, via cooling towers. That, and distribution loses, are major contributors to electricity having a net efficiency of only about 33%.

In Sweden, and other European countries, they pipe the waste heat to nearby, and not so nearby, residential and commercial buildings. It is then called cogeneration of electricity and heat. If a cogeneration plant has excessive heating capacity, they think nothing of sending that extra heat as far away as 15 miles, with a loss of only a couple of degrees in temperature.

The biggest problem with two-way control valves is something called wire draw. Lots of water flow, through a small opening, causes wire-draw. This is where the heating medium actually wears on the plug and seat as though someone was taking a file and "drawing" it back and forth across the opposing closing surfaces until it can no longer close off the flow of heat. It is caused by valves that are too big and heating water that is too hot, forcing the valve to operate in a mostly closed position, with the plug and seat too close together.

You needed to know this to understand that the Swedes have decades of experience in controlling high temperature, high pressure hot water. Again, two-way heating control valves are manufactured, at least, up to 6" by many manufacturers, but not by Tour and Andersson. At Tour and Andersson, they not only make control valves that can control water flow down to 1% of their maximum flow, without causing wire draw, they only make high temperature, high pressure valves up to 2.5". If the situation needs flow greater than what that size valve can provide, you need to install additional valves in parallel, operated in sequence.

In the St. Paul district heating system, I provided several four-valve installations. If I sized the total flow capacity correctly, that means that the water flow could be controlled down to 0.25%, without causing wire draw! After more than 25 years, I've never had a problem call. Again, when wire draw occurs, your occupants will frequently let you know about it, because of overheating, but not always.

As an example of a wire draw problem that may go completely unnoticed, take the sequence of operation of a unit ventilator. If the space is overheated, the thermostat tries to close the heating valve, then open the outside air damper to bring in additional cool air to lower the space temperature. There is no space for a mixed air control, so the outside air is controlled by a discharge control that limits the temperature to a minimum of 55°F to 60°F. If the wire draw damaged control valve is leaking hot water or steam into the coil, the unit vent continues to open the outside air damper for additional cooling, until such time that the minimum discharge temperature is attained. However, because of the leaking valve, it may take a 45°F, or colder, mixed air temperature, to overcome the waste heat from the control valve that can't close. Everyone is comfortable and the wasted heat goes unnoticed, except to the utility company!

I'll talk more about ultrasonic fault detection, in Chapter 3.I. For now, just know that some water, and all steam, leaking through a small opening, make an ultrasonic noise. It can be heard with an ultrasonic fault detector. It's the same device for testing steam traps, and finding air leaks in pneumatic temperature control systems, so it's a very good tool to have. (If you're a school district and your buses have air brakes, you can use it to find leaks there too.) Simply drive the control valve closed and make contact between the valve body and the ultrasonic contact probe. If the valve is leaking steam into the heating coil, you'll know it right away. Again, water leakage doesn't always make an ultrasonic noise, but it's worth trying.

Small two-way valves are throw-aways. Don't waste your time rebuilding ½", ¾" or even most 1" steam valves. If you have problems, just replace them and move on.

In addition to many heating valves being oversized, the other big cause of wire draw on 2-way control valves is simply that the heating water is too hot. Heating water that is too hot also causes the valve to operate in a mostly closed position, most of the time. In steam systems, wire draw occurs even faster. We'll talk in great detail about water that's too hot in Chapter 5.C, but know that, if you've got that problem, I have the solution.

5# steam is 225°F. The temperature cannot be changed. So, on days when a hot water system may be able to take the chill off the building with 100°F water, the steam system is still providing a 225°F source of heat. Wire draw is certain to occur, because the valve has to operate in a mostly closed position, most of the time.

If you're going to keep steam as your heating medium, to help lessen the problem of wire draw, you might want to reanalyze the size of the steam valves. For over 35 years in the industry, when we had enough room, we replaced 1" and ¾" steam valves with one pipe size smaller. If ½" valves were used and seats with smaller flow coefficients were available, we would go down one size. For larger valves, valve packings usually outlast the ability of the plug and seat to close off to the flow of steam, so don't just repack the valve stems, rebuild them with new plugs and seats.

Okay, enough about plugs and seats. It seems that temperature controls are the most tinkered with part of any HVAC system. When something doesn't work, people just tinker with it until they think they've approximated proper operation. The following photos are three good examples:

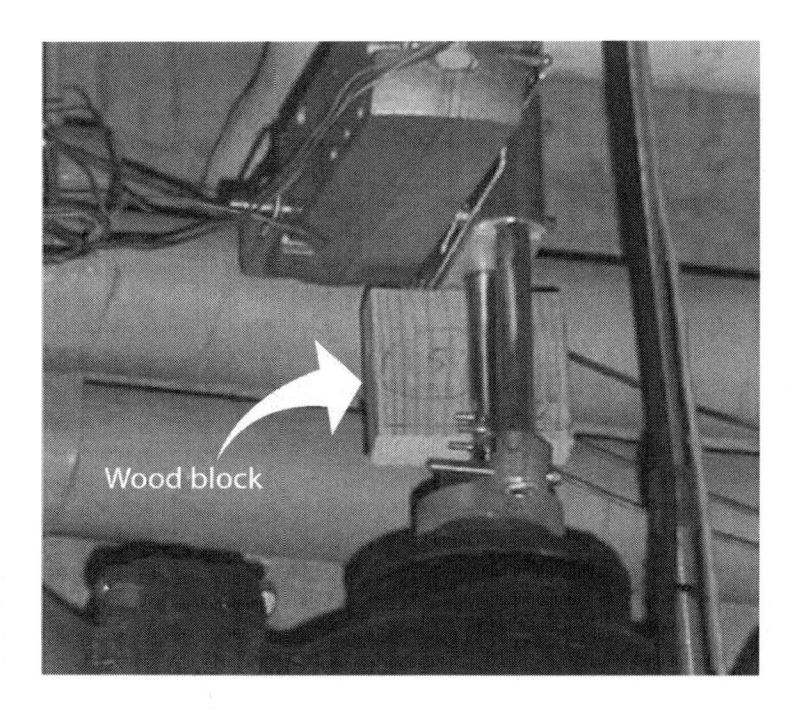

Figure 1.D.1 5" wood block, control valve spacer – one of a series of different sized spacers

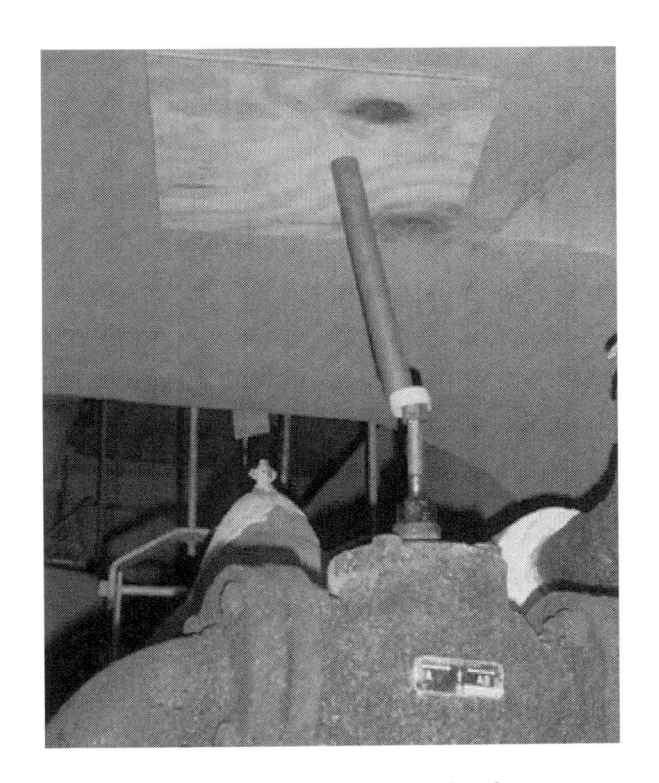

Figure 1.D.2 Copper tube control valve spacer – one of a series of different sized spacers

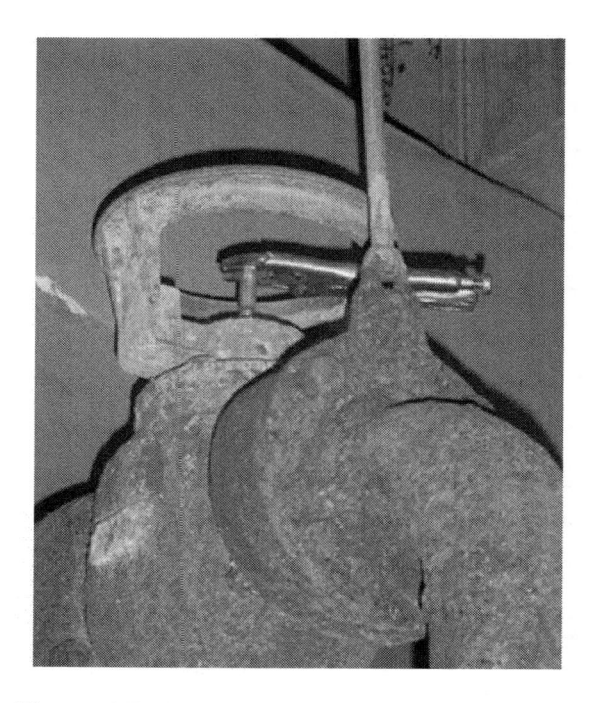

Figure 1.D.3 **Where did I leave my Vice Grip?**

Once again, this is the time to reflect on the life expectancy of steam heated facilities. If the building is going to be around for 15 years, or more, you should investigate converting to hot water heat. Start by calculating a 30% to 50% reduction in heating energy savings then add a 72% reduction in the cost of maintenance, as estimated by one study. That's what you can expect from converting to proper hot water heat. Now, here is the catch. The 72% reduction in maintenance cost is from a good steam system maintenance program to a good hot water system maintenance program. The savings may be less, if the current steam maintenance program is substandard! See Chapter 6.D on benchmarking, to help project your facilities' specific potential savings.

One more thing - if you're converting from steam to hot water heat, and your engineer thinks you may be able to use the old control valves, the issues of oversized valves and wire draw on plugs and seats, may not be as big an issue. With less heat in the hot water, the bigger control valves may be okay. With the ability to lower the temperature of the heating water to whatever is needed, the little bit of heat wasted may not add up to much. The bigger issue may be leaking valve stem packings that will be operating at higher system pressures. For certain, you'll want to hire a mechanical engineer, but don't hesitate suggesting that he (she) make reading this book mandatory.

Fix what's broken

1.E - Steam Boilers

Boiler Safety

The first thing I want to do is direct you to a website provided by Hartford Steam Boilers (HSB): http://www.hsb.com/HSBGroup/uploadedFiles/HSB_COM/Information_Resources/Heating%20Boiler%20Start-Up%20Checklist.pdf.

This particular page is a heating boiler startup checklist. I'd encourage you to print this page and give to each of your chief engineers. You should note the similarity to many of the recommendations in this book. That is, keep everything clean and dry and you'll have a good start on a great maintenance program.

There are lots of things we can do to make the boiler system more efficient and reliable. And, nothing is more important than a good fall tune-up. The very first concern with any boiler burner is safety. The items that Hartford Steam Boilers point out, at their website, are items that must be completed, each heating season, to ensure a safe operating boiler. These safety devices need to be checked by competent, licensed technicians in the boiler/burner industry. Just imagine what might have happened if the low water cutoff problem in the following photo had not been discovered during the first fall tune-up provided by one of my company's servicemen.

Figure 1.E.1 Low water cutoff safety control, with some kind of gunk found during fall tune-up.

The following sign was found on the front of a big box steam boiler. In my PowerPoint presentation, it shows up full-sized. But, when transferred to Word, much was lost, so I ask you to imagine it as a sign.

BOILER IS OFF LINE
TO BRING ON LINE PLEASE DO THE FOLLOWING:

OPEN MAIN STEAM STOP VALVE
OPEN WATER TO THE BOILER – THREE VALVES
TURN ON THE POWER
TURN ON THE FUEL

AND STAND BACK!!!!

Figure 1.E.2 Not exactly something that instills a lot of confidence about the boiler's safe operation.

In addition, a number of these items require daily, weekly or monthly maintenance checks. Who better for the in-house maintenance personnel to learn proper ways to perform these tasks, than from the competent, licensed service technician? Be certain the company you hire knows that you want to build a solid working relationship with them and you want their service technician to help teach your people the proper way to operate and maintain the equipment between professional service calls. Make certain your people learn the correct procedures on those items that it's okay for them to check and what items for which they should wait for the professional.

Good quality, properly calibrated stack thermometers are a must for every boiler. When the boiler is properly tuned, your service company's report should record the stack temperature at the various firing rates. You should take time to note the stack temperature, at least, once every week. From the clean, proper operating condition, a change, up or down, 40°F, should prompt a call to your burner service company.

Combustion Efficiency

Burners need to be tuned, at least, once each heating season. If you are a large fuel user, you may want to have it checked more frequently. A linkage that slips, a linkage rod that is bent, or a dirty primary air blower wheel, can cause excess fuel to be burned.

In Minnesota, most duel fuel boiler burners did not have to be operated on oil for seven consecutive years, prior to the winter of 2013/2014. Boiler/burner service companies were really put to the test, trying to get all the burners to work properly on oil, when the order went out to switch to oil.

Oil nozzles will carbon over and plug the flow, even though you have not burned oil. This happens faster if the oil has been fired to test and then switched back to gas. The oil remaining in the nozzle is heated by the gas combustion and can cause carbon deposits. A conscientious serviceperson will pull and clean the nozzle (nozzles) after it has been test fired and turned back to gas to ensure that, when you really need to burn oil, it will fire, and fire clean, at reasonable air/fuel mixtures. Because you're burning natural gas most frequently, the burner will most likely be optimized on natural gas and burning oil will be a compromise. Again, be certain to clean the nozzle(s) after firing on oil.

Now, I'm also not an oil specialist, but, if you haven't burned oil in some time, you should have the oil checked for potential problems. I often hear of moisture in the oil tanks from condensation. That brings up another point. Many facilities maintain interruptible fuel supplies of thousands of gallons of oil. I was recently looking at a local gas company's website for some current pricing information. There I learned that they now recommend a backup fuel supply of just 20% of the coldest month's consumption. That's a far cry from earlier recommendations, where a modest sized school, may have 5,000 to 10,000 gallons of standby oil. I believe, the larger the backup fuel supply, the more likely you are to have issues.

Combustion air intake

Boilers require 1 square inch of unrestricted combustion air intake for every 5,000 Btu's of input for proper operation. However, when combustion stops, it's a really good idea for the combustion intake to go closed, or you might end up with a snowbank in your otherwise overheated boiler room.

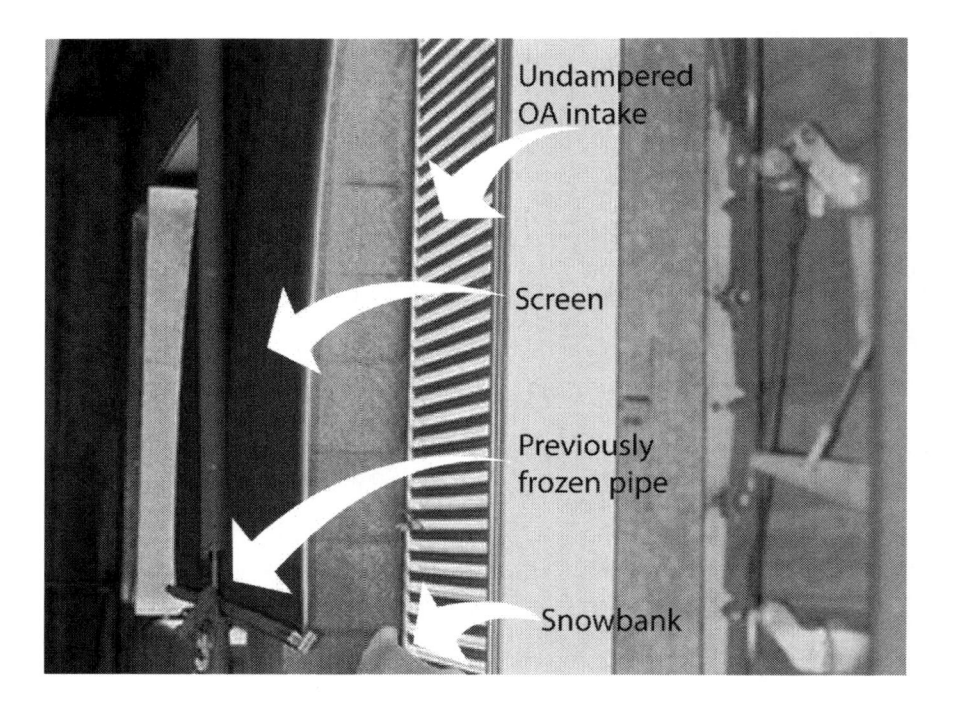

Figure 1.E.3 Undampered combustion air intake, with snowbank at the bottom. Screen installed to help keep pipe from freezing. It didn't work!

Figure 1.E.4 More freeze protection help, because of undampered combustion air intake. Same boiler room.

In 35 years in the industry, I've never seen a ceiling hung destratification fan in the boiler room. This might be a good spot, along with a damper that closes properly.

Leaking fire tubes

Leaky fire tubes are often caused by thermal shock. In steam boilers that may be cold water, and a high flame. In water boilers, it's often the use of a 3-way mixing valve that allows return water to enter the boiler at temperatures more than 30°F colder than the water in the boiler. When the building goes into morning boost, and cold water returns to the boiler too rapidly, thermal shock can often occur. I probably don't need to remind you that re-tubing is a job for professionals. I'll discuss mixing valve operation, in detail, in Chapter 5.C.

General Boiler Issues

I've only presented some basic boiler information that I hope will help. I have not tried to answer all questions about fixing what's broken, when it comes to boilers. Along my career, I have frequently visited a website that has helped me considerably, when it comes to steam and boilers: www.heatinghelp.com.

Its primary author is a fellow by the name of Dan Holohan. While I disagree with his recommendation of the steam trap manufacturer that, apparently, helps sponsor his website, I trust his years of experience. It should be noted that I receive no funding from any of the manufacturers mentioned in this book. Other areas for which you may get additional boiler information from Dan include:

- Flooded steam boiler
- Burner short cycling
- Burner shuts off on low water
- Condensate feed water pump not operating as it should
- Return lines clogged
- Surging
- Foaming
- Water hammer

Not being a mechanic, I was honored, and humbled, when my pipefitter business partner, and exceptionally knowledgeable service foreman, bought me one of Dan's shirts, normally only worn by a mechanic, that says:

A hundred years from now, they will gaze upon my work and marvel at my skills, but never know my name. And that will be good enough for me.

Thanks, Doug!

Conclusion

A well-tuned burner will mix air and fuel, at all stages of capacity, to maximize combustion efficiencies. A good burner tune-up will more than pay for itself in the energy saved. A thorough burner tune-up can prevent costly emergency service calls, at the most inopportune times. In addition, you'll have the peace of mind knowing that your boilers are operating safely.

We can't encourage you enough to get your burners tuned up by a competent, licensed, service technician. Make certain that you get a copy of the combustion analysis, so that changes in operation from one tune-up to the next can be noted. Be certain that you allow, no, **INSIST** that your maintenance personnel participate in the tune-up process, with the tune-up specialist. A good mechanic can be a great instructor. Let your service company know that it's a great way for you to get to know, like and trust them. Remember, most natural gas utilities will provide a rebate for boiler tune-ups.

Finally, if your burner is 15 years old, or greater, you really should consider replacing it. Some of today's modern burners can provide a minimum 5:1 turndown ratio. That means that the boiler can maintain a constant fire at a heating load of just 20% of the design maximum firing rate. This means: fewer on/off operations; fewer costly purge cycles; longer lasting equipment life; etc. A 5:1 turndown ratio is also a requirement to qualify for most gas company burner replacement rebates! New boilers and burners will be discussed further in Chapter 5.B.

Clean what's dirty

2.A – Outside air intakes

We've all heard the saying that cleanliness is next to godliness. If I ever get to heaven, put me in charge of the HVAC system. I'd like to finally see one HVAC system that didn't look like the devil!

Remember the outside air intakes pictured in Chapter 1.C that were either partially or totally covered? I can't tell you the number of times I've found outside air intakes half, or more, covered with tarps, cardboard, plywood or insulation. When asked why, trying to reduce the amount of outside air, is the common response. To paraphrase what Paul Harvey might have said, you need to know the rest of the story.

Behind every rain louver is a bird screen. It is designed to keep birds and other small animals from setting up housekeeping in your facilities. Bird screens come in three sizes. The first, and most preferred size, is a ½"wire cloth. It's a good compromise between keeping all undesirable foreign bodies out and not getting totally plugged in a short period of time. But, even the ½" bird screen will still get plugged. In areas of the country where cottonwood trees spew out their debris, it usually takes about 5 years until nearly all airflow has stopped.

Figure 2.A.1 ½" bird screen that is all but totally plugged with cottonwood lint.

Most unit ventilators have their outside air intake louvers very close to the ground. As the lawns are mowed, everyone puts the left edge of the mower tight up against the wall, blowing the grass clippings, dandelion seeds, cottonwood lint, etc., away from the building. But, when reaching the other end of the building, the mower too often just turns around and blows that same debris right into the outside air intake opening, accelerating the process of plugging the bird screens.

Please instruct the maintenance workers that mow the lawns, to blow the grass clippings and debris away from the building, for at least 10 feet. If the conditions are such that debris still flies against the building, have them move even farther from the building, as far as necessary, to keep from having any debris blown against the building.

How bad can this problem get? My company had performed some temperature control system revisions to all the unit vents in a junior high school. On the first nice spring day, the director of buildings and grounds called to say that the head of his maintenance staff, who considered us his competitor, instead of his ally, was telling all the building heads that our control revisions had eliminated the use of the outside air for cooling. So, I immediately went to check things out.

Having experienced similar situations, I went right to the outside air intake. Fortunately, there were just 4 bolts holding the rain louver in place. When removed, I found the following situation:

Figure 2.A.2 Unit vent bird screen, almost completely plugged with debris blown through the rain louver into the bird screen by lawnmower.

Cottonwood lint and grass clippings were so thick on the bird screen; the debris was in rows, just like the rows on the rain louvers. In the exact center of the bird screen, there were about 6 small holes where the ½" bird screen could be seen. No other portion of the bird screen was visible. Thus, the classroom was getting very little ventilation, or atmospheric cooling.

I took the bird screen away from the building, so the debris wouldn't be pulled right back into the building. Plus, there was a curb about 30' away, against which I could pound the screen to remove the debris. Again, it was about a 60°F day. After I got the bird screen and rain louver reinstalled, I went to the previously severely overheated classroom to find everyone exclaiming how comfortable it was. "What did you do?"

Unfortunately, the chief engineer that was already spreading the news that my company had screwed up, never stepped up to admit his error. It took his retirement before we ever returned to further help that school district.

In another suburban area school district, when finished cleaning the debris from the area between the rain louver and a single unit vent, the in-house personnel were so impressed with the full trash can, they weighed it. It weighed 53# and included a pheasant head!

Figure 2.A.3 Outside air intake, with bird nests.

The outside air intakes of air-handling units usually have lots of air moving through them. No self-respecting bird would be dumb enough to build a nest in a wind tunnel, so if you find something like the above photo, it should be a big red flag that something is wrong. Look for something downstream causing little to no air flow.

Okay, so the bird screens need to be cleaned –regularly! Then, why is it, that so few systems provide any means of access to the bird screens? The rain louvers are often cemented into the brickwork and/or bolted to the inside of the structure before the unit ventilator, or air-handling unit, is installed. Removal without destroying the louvers is nearly impossible. Unfortunately, destroying them is what often has to happen. I estimate that nearly 75% of all unit vent rain louvers have 4 easily removable bolts that hold them in place. The remaining 25% of the unit vent rain louvers will, most likely, have to be destroyed before you get them out. You'll have to work with a sheet metal contractor to design a rain louver that will be able to be removed the next time the bird screen needs to be cleaned.

As you can well imagine, the smaller the mesh of the bird screen, the more frequently it needs to be cleaned. Again, it is essential that ½" bird screens be cleaned every 5 years. ¼" mesh bird screens require cleaning about every 3 years.

Finally, believe it or not, in approximately 2000, we found brand new HVAC systems, factory built, with mosquito screens for bird screens! The devil certainly had his hand in that design! During the very first cottonwood lint season, these screens became totally plugged within just a few days! It was quite a task to remove the louvers from every new air-handling unit, to replace the mosquito screen with ½" wire cloth.

Today, numerous horizontal unit vents (HUV's) are being replaced in large numbers, by vertical unit vents (VUV's). The claim is that HUV's can't deliver the proper amount of ventilation. I wonder how many of those claims are really caused by plugged bird screens, as illustrated here!

Clean what's dirty

2.B – Cleaning Unit Ventilators and Heating and Cooling Coils

Okay, the bird screens have been cleaned. The areas between the bird screen and the unit ventilator, or air-handling unit, have all been cleaned. The people mowing the lawn have all been instructed to blow the debris away from the building for, at least, 10 feet. Now, let's go inside to clean the unit ventilators.

I've been on the outside of schools, where I've seen gray clouds of debris coming from rooms, where air compressors have been used to try to clean the unit vents. The maintenance personnel have placed box fans in the windows above the unit vent, in an effort to exhaust all that debris outside. It would appear that this method of cleaning just causes additional problems.

High-pressure spray washing of unit vent coils is best. If there is a lot of dirt built up inside the unit, vacuum clean the entire unit first. No need to make mud if you don't have to.

Figure 2.B.1 Side panel of dirty unit vent.

Figure 2.B.2 Fan and motor assembly for unit vent.
How many years since cleaned?

Figure 2.B.3 Dirty unit vent and metal washable filter, in great need of cleaning.

Have a tin pan built to fit into the area where the air filter goes. The pan should be wide enough to completely fill the filter space of the narrowest unit ventilator in the building. Measure the front to back dimension of the unit vent filter track. Add 12" to 15" to the front to back dimension of the pan. That way, when inserted into the filter space, it sticks out the front of the unit vent by that much. Of course, the height of the pan should be slightly less than the filter opening height. This pan should be water tight. It is going to collect water and debris from the heating coil.

Start by vacuuming everything possible. Prior to high pressure washing, take precautions to ensure you are not going to get water inside the motor or electrical control box. If the motor is in the air supply, remove the motor and fan assembly, if possible. Some units allow you to easily remove the fan & motor board in one piece. Hopefully, that's the case with your unit vents. Blower wheels are easier to clean, if you can take them outside. Vacuum clean the motor intakes, then wrap the motor with plastic to keep dry. Then pressure wash the dickens out of the fan!

Inspect the unit to determine if there is any way for excess water to get away from you and into the piping tunnel or ceiling space below. Seal all openings with "fire stop" caulking. If the fan has to be left in place, and the individual fan housings do not open to allow for cleaning, you will probably have to drill a ¼" to 3/8" drain hole in the bottom of each section of the fan housing. Cover each hole with a piece of aluminum duct tape, when finished cleaning. Make certain that you have your wet/dry vacuum ready, insert the pan and it's time to start pressure washing the coil.

Back inside, wash the unit from the top down to the pan. Inspect the coil with a flashlight to make certain that you get it clean. It will be helpful to have a second person to run a wet/dry vacuum, just in case the pan gets full before everything is clean. There will inevitably be some water that splashes on the floor too. Vacuum it up right away.

The area below the filter, and through the damper section, has to be thoroughly dried immediately after washing. This is very important to avoid mold growth. When finished washing, use a rug-drying blower to dry the unit.

Now would be a good time to lubricate the motor, bearings and damper linkages. Be careful to not over lubricate. This causes dirt to adhere & the odors can alarm the occupants, even when they are harmless.

If the unit has a belt driven fan, now's the time to change belts. Don't wait until you have to make a special trip, just to change a belt. New belts are cheap!

When finished cleaning the next unit ventilator and you're returning to get the rug-drying fan from the previous unit, put in a clean filter and start the fan. Run the fan until you're comfortable that the unit is completely dry.

On one memorable unit vent cleaning occasion, the two-person team consisted of one of our servicemen and one in-house person. Having firmly established a great working relationship, I made regular sales calls to their facility. On one such call, about 5 years after the coil cleaning project, the same in-house person told me he had decided to clean the unit vents, again. After similarly washing just one unit, he put all the tools away, stating that the wash water was still clean. That in-house person was thrilled to see the results of his good filter maintenance program, which will be discussed next.

Clean what's dirty

2.C - Unit Ventilator Air Filters

Most unit vents are found in schools, but there may be similar units in applications like hospitals. There may also be similar filter applications for ceiling mounted fan coils.

Life is full of compromises and unit vent filters are no exception. Use a filter that stops most of the dirt and it also stops most of the airflow. Use a filter that allows nearly 100% air flow and you'll be pressure washing your coils every year. My simple rule is, **"if you can read a newspaper through your filter, don't use it."** This applies to your home filters too.

The first key to any good filter is its ability to depth-load debris. That is, the big chunks should collect on the course entering side. Smaller chunks then collect on the ever denser, and perhaps tacified media, as the air passes through the filter.

Another key to a good filter system is that all the air has to pass through the filter. That is, insist on a system where there is a good seal around the edges of the filter.

One of the difficulties with unit vent filters is that there are no standard sizes. In a typical school, you could easily have a half dozen different sized unit vents. In the same district, you could have dozens of different sized filters. So, there are no standard sized cardboard frame filters. They are all special order sizes and quite expensive. In addition, they don't work very well.

The unit vent filter system I liked best may be difficult to duplicate elsewhere in the US. I will describe the filter system I found worked better than anything else, and hope you can find someone in your area that provides a similar, or better, system.

Whether fiberglass, polyester, or a layer of each, a two layered **roll** media worked best for us. The two layers are sewn together, with about an inch of material on the outside of the seam. That extra material helps keep all the air flow through the filter and not around it. If you have an appropriate location, the roll media can be hung on the wall, like a roll of toilet paper. Place a paper cutter on the table below the roll, and pull the media edge to a piece of tape, marking the appropriate length of the filter. Whack it off with the paper cutter and repeat.

The manufacturer that we used had an option that the leaving layer was what they called a scrim backing. The scrim backing was a tightly woven synthetic fabric that quickly plugged, stopping virtually all the air flow, in a matter of a couple of weeks. Do not use filters with a scrim backing.

The filters are slipped over a wire, or copper tube frame, that should be sized for each unit ventilator. The frame is made about ½" shorter than the filter space measurement, from side to side and about 1" shorter from front to back. The seam to seam dimension should be about ½" wider than the frame, so the filter slips easily over the frame.

When changing filters, always know what size filter you need for every location. Put this list in a plastic folder and put it on the filter cart, along with a large garbage can, lined with a big plastic bag. Simply slide the dirty filter off the frame, directly into the garbage can. Slide the new filter on the frame and reinstall.

Always take a quick look at the area below the filter. If it's loaded with debris, take a minute to vacuum it up. If you don't, the filter may need changing in just a few days. You might want to also ask the custodial staff to be certain that the unit vents are off, when cleaning the floor. This will keep loose airborne debris from being drawn into the filter, shortening its life.

Speaking of frequency of filter changes, unit ventilators in schools need changing, at least, 4 times a year. The first time needs to be by November 1st. My experience has been that, at least, 50% of my company's service calls, after November 1st, are dirt related. I suggest that you **never** call your HVAC service company, for a comfort complaint, until you have checked to see if the filters are plugged, the coil is dirty, or that the fan belt needs replacing. It can save you a lot of unnecessary service charges.

Christmas vacation is the next time for changing unit vent filters, followed by a change over Easter vacation or spring break. The final filter change is then done during the summer months.

Now, a note about antimicrobial filter treatment. It makes for good public relations, but that's about all. I've been told that the shelf life of antimicrobial treatment is only about 4 months. When you're changing filters every 3 months, it will drive you crazy trying to keep fresh filters on the shelf.

ASHRAE, The American Society of Heating, Refrigeration and Air Conditioning Engineers, states that antimicrobial treatment has no beneficial effect. The air simply is not in contact with the antimicrobial long enough to kill any airborne bacteria. In addition, the debris that is in contact with the antimicrobial acts as an insulator between the antimicrobial and any additional debris that is not in direct contact.

In cases where a student, or teacher, is having health challenges, you might consider spraying an approved antimicrobial into the return air intake of the unit ventilator 2 or 3 times a week. This should kill the bacteria that have accumulated on the filter. Note that it is important to use an antimicrobial that is safe and leaves no trace after just a few minutes. Anything that lasts longer may allow bacteria to build up immunity.

You might also try spraying the same, safe antimicrobial on the pile of toys in the kindergarten room. Kick them around and spray them again. I've been told that lots of childcare centers have seen fewer colds being passed around when they do this.

The filter material we used was by AirGuard. It consisted of a layer of fiberglass and a layer of polyester. At the time, AirGuard made wire frames by simply taking two L-shaped pieces of heavy wire, spot welded them together and used the welder to cut off the excess. Unfortunately, each end had a rough edge sticking out that caught the filter material, as it was slipped into the sleeve. Being a pneumatic temperature control service company, our servicemen knew how to easily bend copper tubing into smooth, one piece frames. That worked great for us, but it may not be able to be repeated elsewhere. It could be that AirGuard, or someone else, has figured out a better system. If anyone has a good unit vent air filter system that has national distribution, please forward their contact information to me and I'll check it out.

AirGuard Filters
100 River Ridge Circle
Jeffersonville, IN 47130
1-866-247-4827
mailbag@airguard.com
www.airguard.com

Clean what's dirty

2.D - Air-handling units, plus heating and cooling coils

Now that the unit ventilators have all been cleaned, and new filters installed, let's start cleaning the air-handling units. Again, begin by vacuuming as much of the surface debris as possible. That includes as much of the return air and mixed air chambers, coils, fans and as much of the ductwork that can be reached. I have recently had the occasion to inspect return air chambers of a number of 28,000 cfm capacity rooftop air-handling units. On many of these units, the debris has accumulated to as much as 1.5" thick. It is imperative that this debris be vacuumed before it breaks loose and prematurely loads your filters.

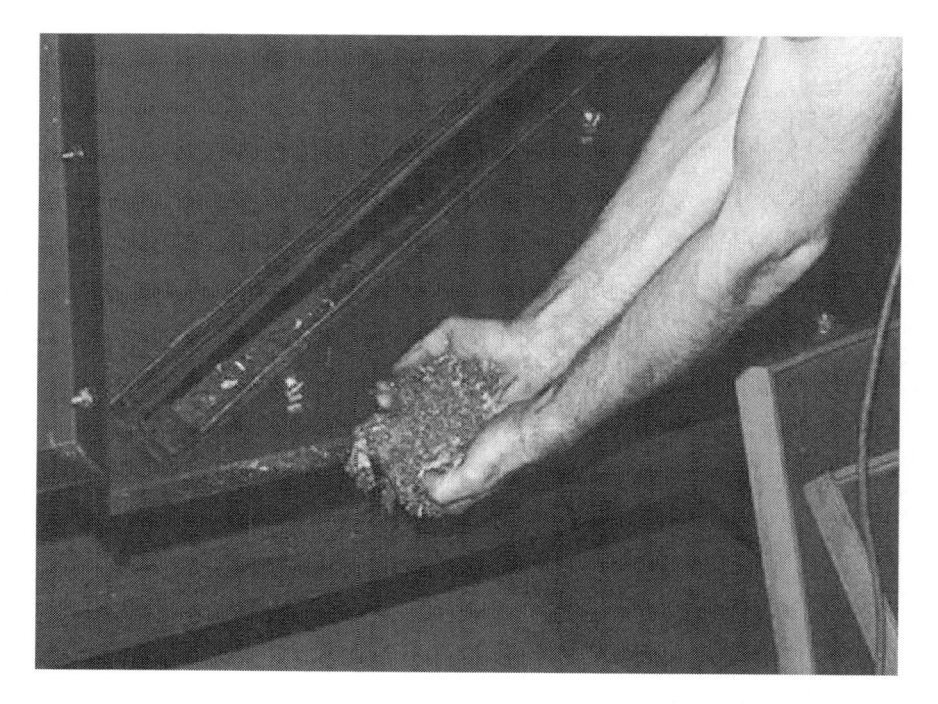

Figure 2.D.1 Mixed air chamber, with bugs and other debris ½" thick on MA floor.

On the above unit, the maintenance department blindly changed the not-so-good filters stacked up on the right. Because of all the debris left on the MA floor, the turbulence in the area has the possibility of churning up all that debris, making it necessary to change the filters within just a few days. Unfortunately, no one was back to change the filters for three months!

Next, and often the most difficult, is to gain proper access to the heating and cooling coils. Sometimes the air-handling unit has to be half torn apart to get at both sides of the coils. High-pressure spray washers can develop pressures in excess of 2,000 psi. For coil cleaning, pressure of approximately 500 psi should be sufficient, as well as safer. Coil fins are often quite thin and very susceptible to being flattened by the spray from the washer that may be directed at an angle to the coil fins. Flattened fins are worse than dirty fins.

The spray nozzles from pressure washers are often designed with a narrow fan styled spray. Open your hand and spread your fingers. It looks kind of like that. If you spray the coil with the fan going cross-wise to the fins, you may flatten the fins. Instead, the spray fan should be parallel to, and straight into the fins. This might slow the perception of getting the coils clean, but I believe, in the long run, it will take less time and you'll do a better job.

Some coils are so dirty that it is recommended to vacuum the surface dirt first. Again, no sense in making mud, if you don't have to. Using a coil cleaner, sprayed on with the pressure washer, or with a Hudson sprayer, is often a good idea. Your local wholesaler, or Goodway (See note at the end.), should be able to help you with this.

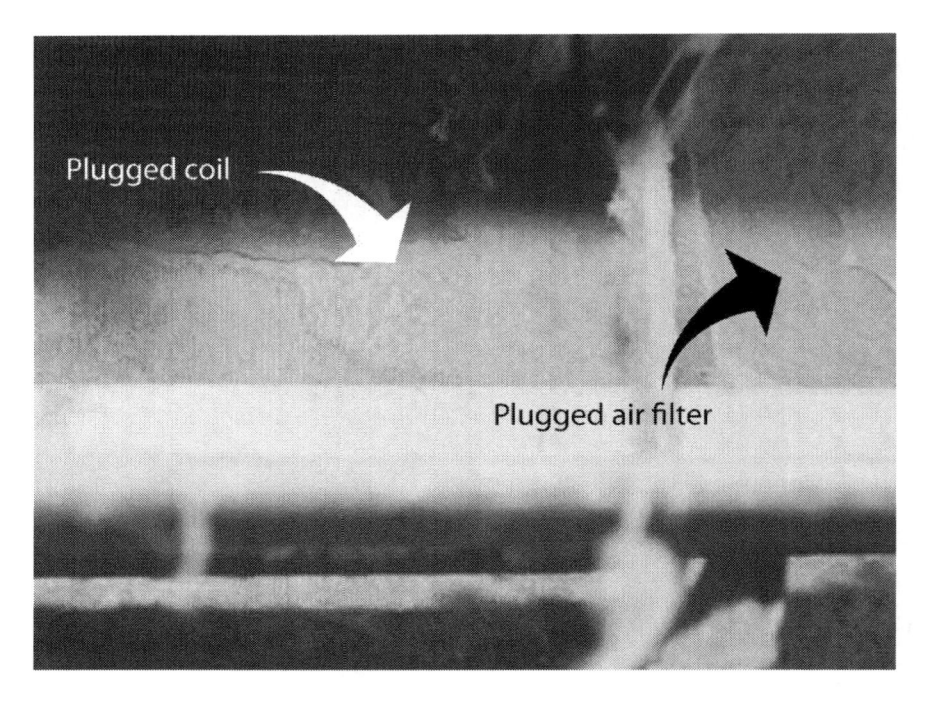

**Figure 2.D.2 Dirty coil and neglected filters.
Not that unusual.**

In the above photo, you can see where the pressure washing, from the back side had begun. The debris was literally rolling off the coil. This is probably one of those situations where vacuuming should have been done first. On the right side of the photo is the standard bottle cap style fiberglass filter. They are not very effective. It has been far too long since these filters have been changed.

When I first started sending the paper form of my monthly information, before the Internet existed, it caught the attention of the in-house architect for one of the largest school districts in Minnesota. They were having trouble heating one large classroom, heated and ventilated by an air-handling unit that only served that one room. They asked me to investigate – for free, of course. Okay, the district architect can't figure this out, so I was quite excited to finally use my engineering skills. In the hope that it would lead to doing some business, I agreed.

When I arrived, the maintenance engineer first took me to the classroom. Other than a lack of airflow, there wasn't anything to learn there. I asked to see the air-handling unit. On the way, I asked him to tell me about the filter maintenance program. Each month, he would get a stack of postcard sized maintenance tasks, including a card to check the filters, quarterly. When opening an air-handling unit, I try to make an effort to turn the unit off, hoping to not disturb the debris collected on the filters. On this occasion, I forgot and just unlatched the door. Usually there is suction on the door, but once unlatched, the door freely swung open. That told me there was no airflow.

I then pulled out the standard bottle cap style fiberglass filter, that I could have read the paper through. It was perfectly clean. When I asked the maintenance man how often he has to change that filter, he said that he had only been in that facility for four years and that was the same filter that was there when he started. Filters should show signs of collecting debris in a couple of weeks. A filter that hasn't collected debris in four years should be screaming that something is wrong!

I took out the filter and shined my flashlight on the heating coil location, about 15" downstream of the filter. I intentionally said heating coil location, instead of heating coil, because the heating coil was not visible. The debris was so thick, I pealed back one corner, then another, allowing me to peel off a one piece, ½" layer of debris that completely covered the coil. It looked like an old, green Army blanket! The maintenance man was shocked and embarrassed.

Before leaving, I asked if I could have a look around the boiler room. I can often find issues with the pneumatic control compressor system, and more, just walking through. Yes, there was an opportunity to help assure a clean, dry, oil-free source of air for the controls. But, what really surprised me was the flume of steam coming out of the condensate tank. Instead of repairing or replacing the defective steam traps to solve the problem of condensate collecting on the floor below the vent, they simply added a couple lengths of pipe to get the flume closer to the floor drain.

So, now it's time to give my report to the architect. It turned out he had not looked for the problem. He just asked me to find it. But now, it was time for in-house politics to enter the fray. It turned out that this district also had a licensed mechanical engineer in charge of mechanical maintenance. He was ticked that the architect was investigating an HVAC issue. As if that wasn't politics enough, the district also had a licensed mechanical engineer that was in charge of energy conservation. He was now ticked that the architect was looking into the energy savings that would come from a steam trap repair program. Do I really need to tell you that my company never turned a wrench out of all this?

Okay, let's get back on topic. Start by pressure washing each coil in the opposite direction of the airflow. This helps drive the debris out the same way it came in. With really thick coils, it is sometimes necessary to wash coils, alternatively, against and with the airflow. You may have to repeat this procedure several times, until the water runs clear. Be careful to not damage the interior insulation. Also keep in mind how long it took to gain proper access. Once started, don't quit until you're certain that everything is properly cleaned.

If you can't vacuum the wash water out of the bottom of the fan housings, drill one or more 3/8" holes in the bottom of the fan housing to allow the water and debris to drain away while cleaning the blowers. Cover the holes with aluminum duct tape when done.

Most heating only units do not have drains for the water and debris to flow to, so it is especially important to have a second team member with a wet/dry vacuum. Be certain to keep the water and debris concentrated in the air-handling unit, so you do not have to chase water all over the equipment rooms.

Ceiling hung air-handling units, and some horizontal heating coil air-handling units pose special difficulties in water control. We have found that taping a sheet of plastic all the way around the bottom of the air-handling unit, and making it into a funnel, which directs all of the water and debris into a garbage can on the floor, to work the best.

After cleaning the entire unit, lubricate all bearings and damper linkages. It's a good idea to spray the linkages during unoccupied times. I know of schools that have been evacuated when a teacher smelled WD-40 coming out of the diffusers.

Install good quality air filters and run the fan to thoroughly dry the interior of the unit. Again, this is best during unoccupied hours, to prevent IAQ problems or complaints.

For more information on coil cleaning, coil cleaning equipment and chemicals, Check the info at the following website:

http://www.goodway.com/products/coil-cleaning-systems-chemicals/coil-cleaning-machines/coilpro-cc-400hf-hiflow-coil-cleaner#.U4clGItOU2w

Clean what's dirty

2. E – Air-handling unit filters

How are you doing on your air-handling unit and coil cleaning efforts? What a dirty, nasty job! Wouldn't that be a good task to do as infrequently as possible? It's all up to the quality of air filter material you use and the frequency with which the filters are changed.

I talked extensively about the unit ventilator filters previously. Some of the same rules apply to air-handling system filters:

1. Make certain that you can't read a newspaper through it.
2. Make certain that it's capable of depth loading the big chunks on the entering side, with smaller dust particles being caught by a denser, tacified layer on the leaving side.
3. Make certain that all of the air goes through the filters, not between or around them, as with cardboard, plastic or metal-framed filters.

Figure 2.E.1 Example of the debris allowed to flow between and around cardboard, plastic or metal frame filters.

For the most part, the filters used in the above photo do a pretty good job of cleaning the air that goes through the filters. The problem is that not all the air goes through the filters. The debris on the left side indicates that the filter isn't extending all the way across the rack. I'm guessing someone was too lazy to put the rack filler back in place after changing the filters. A common occurrence.

The middle two accumulations of debris indicate the inability of the filters to keep airflow from going between the filters. The debris on the far end of the coil is from the filter not sealing tight against the far door.

To accomplish proper filtration, I recommend the use of a "linked" filter. An internal wire frame supports two layers of polyester material, heat sealed around the frame. The links hold the filters together, not allowing any flow of debris between the filters. You provide the manufacturer with the exact filter track sizes, top to bottom and side to side; the number of rack channels per air-handling unit; and the name or number of the unit. The filters are then shipped in boxes, properly tagged with the unit identification. When removed from the box, the individual, mostly standard sized filters, are linked together, so you can just push them all the way to the other door. When the filter access door is closed, the filters seal on both ends of the unit. And, when the blower is on, the air pushes the tacified leaving layer against the track, again keeping the air from blowing around the filters. This one piece assembly practically eliminates all ability for air to bypass the filter. To remove the old filters, just pull on the outside filter and the rest follow.

The filters I recommend do contain an antimicrobial. As described previously, many sources aren't too keen on filters with an antimicrobial. However, the features of this filter are so superior that it's still the best filter for the price that I've ever found. These filters also make excellent pre-filters for air-handling units with bag filters.

Figure 2.E.2 One-piece air filters struggling to stay in place.

This photo of the filters I just described, illustrates the course layer of white polyester that collects the bulk of the big chunks. The green side is the tacified layer. If the bow to the filters doesn't tell you that the filters are way past time to be changed, look at the visible discoloration on the leaving green side. Even if you have monometers measuring the pressure drop across the filters, you need to visually inspect the filters, at least, midway between changes. Filters will sometimes collapse under the pressure. If you're just looking at the monometer, you may never know that something is wrong.

How often should you change filters? Ideally, just before the pressure differential on the filter starts to pull the debris out the leaving side of the filter. Most filter manufacturers' literature states that they can withstand a certain pressure differential, often 0.5" of water column. Check your manufacturer's literature to be certain. Just remember, these pressure differentials are measured in ideal laboratory conditions. We don't have that in real life HVAC systems. If you have pressure differential gauges across the filters, watch for discoloration on the leaving side of the filter and note at what pressure differential that occurs. The discoloration is a sign that it is starting to unload. However, never rely on pressure differential gauges alone. Again, there needs to be periodic visual inspections, to make certain that the filters haven't collapsed. Also, make certain that all pressure differential gauges go to zero when the air-handling unit is off. If you're going to rely on them, they have to work properly.

Okay, I understand that what I just described is ideal. In actual practice, all classroom air-handling units should have their filters changed a minimum of 4 times a year. For administrative area air-handling units, or year-round office applications, I encourage you to change filters a minimum of 6 times a year. If your periodic visual inspections show discoloration on the leaving side of the filter, don't hesitate to change the filters more frequently. Just remember the dirty, nasty job it is to clean those coils. How frequently you have to perform that crumby job is completely dependent on the filter maintenance program.

I briefly mentioned bag filters. These are very expensive filters that really capture small dust particles. If your engineer specified air-handling units with these style filters, he/she probably had good reason. Like most filter systems, however, there is a tendency to try to make them last longer than they should. Look for the same discoloration on the leaving side of the filter to determine when changing is necessary.

Speaking of efforts to make filters last longer than they should; I once found a major metropolitan school district that was having difficulty heating much of the school. Upon investigating, it was discovered that the reheat coils throughout the building had up to 2" of debris on them. The initial problem was that, either the engineer didn't specify access panels on both sides of the coils, or the sheet metal contractor got by without performing the work as specified. So, access holes had to be cut; coils cleaned as good as possible; and covers installed over the holes. This had to be done over 200 times!

Well, how could something like that happen? I had to find out, so it didn't happen again. It turned out that these air-handling units were equipped with automatic roll filters. Roll filters are roughly 50' long and come in varying widths. Automatic pressure differential controls advance the roll, as it senses the filter getting dirty. The speed of advancement, however, usually comes as a shock to the budget, so the automatic advance is turned off. Not being advanced frequently enough, what I usually see is a sagging of the roll filter from the weight of debris collected on it. In this case, the maintenance personnel had been instructed to turn the roll filter around and run it through again! Of course, all that did was to unload the already collected debris. How do you suppose those cost savings compared to the cost of cleaning all those coils?

Scan Air Filter
Minneapolis, MN 55404
612-729-2020
http://scanairfilters.com

Clean what's dirty

2.F – Condenser Coils

As you can tell by the sheer volume of content in this chapter, "Cleaning what's dirty" takes a pretty high priority. I've tried detailing how to properly vacuum and pressure wash unit vents, air-handling units, heating and cooling coils. But, the fact of the matter is, all of these devices have, or should have, some level of protection against getting dirty, with an air filtering system.

With air-conditioning the indoor heat is transferred outside. In 1973, The Educational Facilities Laboratories, in their booklet entitled, <u>The Economy of Energy Conservation in Educational Facilities</u>, stated that a small layer of debris can affect heat transfer efficiency by 25%. Without any air filtering systems, a small layer of cottonwood lint, grass clippings, dandelion seeds, and other debris, can easily collect on air-cooled condenser coils in a single cooling season. If you're not currently cleaning your air-cooled condenser coils annually, I respectfully suggest that you're wasting an awful lot of energy; probably not providing the desired comfort on really hot days; and prematurely wearing out your compressors, well before their time.

Figure 2.F.1 At first glance, this air-cooled condenser looks perfectly fine.

Figure 2.F.2 Under the cover, it's a different story.

The light colored areas of this condenser are where the last rain storm washed off some of the accumulated cottonwood lint. The dark areas are where you can literally grab handfuls of cottonwood lint. I think that's more than "a small layer of debris". When coils are this dirty, it is often best to vacuum the debris first. If you blow or wash the debris out, it is likely to be quickly drawn back into the coil.

Figure 2.F.3 Air-cooled condenser after a lifetime of hail storms.

In Figure 2.F.1, the coil is protected from hail damage by both a cover and by having the coils slanted in towards the middle of the unit, as shown in Figure 2.F.2. But, the coil in Figure2.F.3 is fully exposed to the ravages of hail. (I'd love to tell you the name of the nationally known company, on whose roof this unit resided, but I'm too much of a gentleman to say.) That's something to remember the next time you need to buy an air-conditioning unit. This coil is beyond recovery. However, if the owner had taken a little time after each hail storm, it might have been able to have an extended life, as well as operated more efficiently for all the years between.

Figure 2.F.4 Another condensing coil, with more than just "a small layer of debris".

I'd like to say that the above photos are the exception, but they are, in fact, the rule. On this unit, you can only see a few square inches of the actual coil. This coil is horizontal. The vertical boards behind the condenser are part of a fence. With the space between boards, it allows plenty of room for air to be drawn through the coil by the fans above the coil.

I once found a similar condenser, located on a roof. In this case, the architect didn't want anyone seeing any part of the ugly condenser on his beautiful building, so it was totally surrounded by 8' high walls that went all the way to the roof. There were only small openings, near the roof, to allow the rain to drain out. As the air was discharged, vertically, through the top of the condenser coil, it was simply recaptured by the air trying to come down between the wall and the condenser coil. So, the temperature of the condenser air just kept going up.

As the indoor air conditions worsened, in desperation, the in-house personnel placed Rainbird lawn sprinklers on opposing corners of the walls, high above the condenser. By the time I arrived, the calcium deposits were hanging thick from the bottom of the coil. We tried using the strongest chemicals we dared, but to no avail. The building owner didn't like the idea of redoing the wall to allow for airflow and replacing the condenser coil, so I don't know what ever happened. If they ever got cooling, I'm certain they had to eventually do what we had proposed.

The major electric utility serving Minnesota and Wisconsin has what they call a Saver Switch. In order to limit the amount of purchased electricity on hot days, they turn residential air-conditioning units off for 15 minutes at a time. Unfortunately, residential air-conditioning cooling and condensing coils probably look worse than the commercial coils you've been looking at in these photos. If every heating and cooling coil, as well as air-cooled condenser coil started the cooling season in a clean condition, we wouldn't need a Saver Switch. In fact, I'm convinced that if we all cleaned the air-cooled condensers on our refrigerators, on the same day, the electric companies would be able to register the difference on the amount of electricity they needed to generate. I've tried convincing that electric company to have some public address messages on TV, promoting that as an Earth Day event, but it never happened. Maybe you and I can start that movement right here!

There are two schools of thought when it comes to cleaning air cooled condensers:

1. Use compressed air or CO_2
2. Use high pressure spray washing

Generally speaking, only the smallest of air-cooled condensers should ever be considered for cleaning with compressed CO_2. And, that only applies to those units that are too remote to be easily cleaned by other methods. To do a thorough job, it simply takes too many canisters of CO_2. As with all coil cleaning tasks, the debris is best eliminated by blowing, or washing, in the opposite direction from which it entered the coil.

Whether by air, CO_2 or water, care must be taken to blow/wash parallel to coil fins. If you don't, fins can easily be flattened by any of these methods, decreasing the heat transfer efficiency of the coil, rather than increasing it.

For many of the condenser coils, for which my company was responsible to keep clean, compressed air was the preferred method. It's quick and pretty thorough. Setup time is the biggest deterrent. You either need a few really big units to make it worthwhile, or lots and lots of smaller units. First thing in the spring, and/or at the end of the cottonwood lint season, we rented a 100 HP air compressor and just moved it from one building to the next.

There are three specific instances when cleaning with compressed air is not the preferred method. The first is, if your condenser coils are so plugged that even lots of high velocity air can't penetrate the debris. You'll sometimes get a portion of the debris out with compressed air, and that will definitely speed the process, but these coils often take some sort of chemicals to loosen and emulsify the debris.

The second condenser coil type, for which compressed air isn't very effective is what's called split condensing coils. This is where one coil is applied over another coil, spooning the first coil, if you will, creating a pocket between the two coils. You may even find that the outside coil doesn't look at all dirty, but the layer of debris between the coils greatly affects the air flow, thus the heat transfer efficiency, through both coils.

To gain access to these coils, the entire top of the condensing coil, which includes the condenser fan, has to be removed. Then, the two condensing coils need to be carefully separated, first pressure washing one coil, again, always counter flow to the normal direction of the airflow, then the other coil. Be certain to not have the debris from the second coil spaying on the coil that has already been cleaned. You will most likely need to use a coil cleaning agent with this process. If you have a choice, avoid this type of condensing coils when buying new units.

Another type of coil that can be tricky to clean is the lanced fin coil. This is, generally, a single row coil, but it has little serrated cuts in the aluminum fin. While this increases the coil heat transfer efficiency, it creates little edges that really like to grab and hold cottonwood lint and other debris. It is often best, if possible, to always use compressed air first, and then use a chemical agent, if required, to get the remaining debris. Again, I'd avoid buying this type of condenser coils.

If the surface of the coil looks clean, inspect more than just the surface. The outside does not always tell the entire story. With a pocket screw driver, CAREFULLY open a small area of fin to get a better view between the fins. If you see any debris inside, it should be cleaned. Then, carefully bend the fin(s) back in place.

A good indicator of dirty coils is high suction on the unit service door between the coil and the fan, if there is one. If you cannot readily open the access door, with the condenser fan(s) operating, there is a good chance the condenser coil is plugged.

Another good indicator of a dirty coil, on a vertical discharge condenser fan, is to put your hand directly over the top of the fan. If the discharge seems minimal, or even downward into the top of the fan, try bringing your hand up the side of the condenser. As your hand clears the top of the condenser, if there seems to be a sideways discharge of air off the top of the condenser, the coil is likely plugged.

If all other inspection methods fail, just know that, if your area has cottonwood trees, every air-cooled condenser coil requires **ANNUAL** cleaning! Keeping systems clean, on a regular basis, requires much less work than recovering from years of neglect.

Finally, you just can't be in a hurry cleaning air-cooled condenser coils. I can't tell you the number of times a new mechanic has called in, stating that he had just washed the condenser coil, but that the head pressure on the unit was still too high. Told to wash it again, he calls a second time to state that the system head pressure is still too high. Clean it again! Finally, after about the fourth washing and rinsing, with a reduced system head pressure, the now amazed new mechanic calls in to tell the service manager that he wouldn't believe how dirty that coil was. Oh yes he would! Remember, he was the guy that said to keep washing the condenser coil until the head pressure finally comes down.

Dirty condensing coils, equals high head pressure. High head pressure, equals more work being unnecessarily done. More work being done, equals higher cost of operation and longer equipment run times. Longer equipment run time equals shorter equipment life. Shorter equipment life equals more frequent replacement costs. It's a vicious circle.

Remember, 75% of all HVAC service is simply keeping systems clean, dry and lubricated. No one can perform those services more cost-effectively than well-trained in-house personnel. But occasionally, even the best in-house personnel need help on some of these, as well as more technical tasks. Don't hesitate to call the qualified contractors, with whom you have a know, like and trust relationship. Working together, your maintenance personnel will continue to learn more and more.

For more information on coil cleaning, coil cleaning equipment and chemicals, I offer the following website:

http://www.goodway.com/products/coil-cleaning-systems-chemicals/coil-cleaning-machines/coilpro-cc-400hf-hiflow-coil-cleaner#.U4clGltOU2w

Clean what's dirty

2.G – Boilers

Steam boiler blow down

Hot water boilers are closed loop systems. That is, the water, glycol and chemicals in the heating system, stay in the heating system. Steam boilers are open loop, meaning they are subject to losing water and chemicals, which must be replaced. Because new water has dissolved solids, those solids are left behind in the boiler, when the water is heated to steam. Those solids must be removed on a regular basis. All steam boilers have to deal with solids. This big box steam boiler discussion, should get you started in the right direction.

The first anti-scaling preventative measure is to supply good quality demineralized water as make–up feed water. That said, few commercial buildings ever use demineralized make-up water in their steam heating systems. That's often reserved for large industrial applications.

The solids are usually removed by way of a single blow down valve, most frequently at the back of the boiler. You'll need to work with your boiler chemical specialist to determine the frequency that blow down needs to occur. One blow down, however, is usually inadequate. To get the best results from any blow down, it must have a quick opening, 90 degree turn ball valve. When performing a blow down, the valve must be opened and closed quickly. This gets the boiler water bouncing, in hopes of loosening some of the accumulated debris, so that it can be flushed out. Seeing lots of debris in the blown down is a good thing. As long as you're seeing debris in the blow down, keep opening and closing that 90 degree ball at 3 to 5 second intervals, or as instructed by your boiler chemical company. Still, at best, you're going to do an inadequate job with a single blow down point. If you've ever had the job of cleaning out the mud legs, in all four corners of the boiler, you'll agree that there is usually lots of debris left behind.

Done correctly, each mud leg should have its own blow down. Next to each cleanout, you should find a capped pipe to which individual blowdowns could be installed. The piping should be done with the use of unions, so they can be easily removed for summertime inspection and cleaning. The blow downs on each side of the boiler can be piped together to the common drain, again, usually behind the boiler. Each 90 degree turn blow down ball valve should be opened and closed quickly to get the water and debris bouncing. Now, you may not have to exercise every blow down, at every occurrence. You may be able to simply work your way around the boiler, at timely intervals.

Again, work closely with your boiler chemical provider to work out a good blow down schedule.

Fire tubes

When using natural gas, a properly tuned boiler will create little soot on the inside of the tubes, but when operating on interruptible oil, with burners less accurately tuned to oil, the possibility of soot buildup on the inside of the tubes is a definite possibility.

A 1/8" layer of soot on the inside of the tubes is equivalent to a 5/8" layer of asbestos on the water side. Plus, a substantial layer can build up in as little as only two weeks, resulting in a heat loss of 47%, with an increased fuel consumption need of 8½%.

Water tubes

In 35 years in the industry, I've never seen a water tube steam boiler used in the HVAC industry, but I suppose it's possible. Not having any experience with water tube boilers, I can only tell you that keeping the dissolved solids at a minimum may be even more important in water tube steam boilers. You'll need to talk to your boiler chemical provider and boiler service company contractor to establish a proper blow down system.

Boiler condensation

The products of combustion are carbon dioxide and water. When the temperature of the water in any boiler is below the dewpoint of the water in the burning gas, the water is certain to condense on the cold surface. Only today's modern, condensing boilers are designed to deal with that condensation. So, what is the dewpoint of natural gas? Well, it's 131.3°F. When I first got into the HVAC industry, 135°F was the recommended minimum water temperature. Given the reaction time for many burners to react to a start signal, we soon increased the minimum water temperature to 145°F. I'm not certain what circumstances occurred, but most boilers my company maintained were, eventually, set to a minimum water temperature of 160°F.

As boilers warm from a cold start, there is going to be some condensation. It will evaporate, but you do need to exceed the dewpoint to eliminate it. There are obvious problems created by the condensation itself, but when the condensation eventually evaporates, it will also leave residue, with which you have to deal. It's best to avoid the problem with proper minimum boiler water temperatures.

Regularly scheduled tune-ups are the best way to keep from having to clean the tubes. If you've done it, you know it's a good job to avoid. In doing Internet research, on all kinds of HVAC equipment cleaning, the name that keeps coming up most frequently is:

Goodway Technologies Corporation
420 West Avenue
Stamford, CT 06902-6384 USA

Tel: 203-359-4708
Toll Free: 1-800-333-7467
Fax: 203-359-9601

Email: goodway@goodway.com

I'm certain they can help direct you to the proper methods, and equipment, for cleaning almost everything HVAC related.

Clean what's dirty

2.H - Boiler and equipment rooms

The last items on our list of cleaning what's dirty are the boiler room and equipment rooms themselves. After over 35 years in the industry, I can tell you it doesn't take long to make an assessment of the general condition of any facility. Remember how you've always heard that the first impression is a lasting impression. Well, you never get a second chance to make a good first impression. When the facility owner, corporation president, school superintendent, board members, or others, enter your facility, you all work very hard at making certain of a good first impression. But, what do you suppose they will remember most, if they go in the boiler room or one of the equipment rooms?

Start by reclaiming your equipment rooms. Many of these rooms have been taken over as storage rooms. They are not storage rooms. Get that old junk out of there! Add some lights! Make it a pleasant place to work.

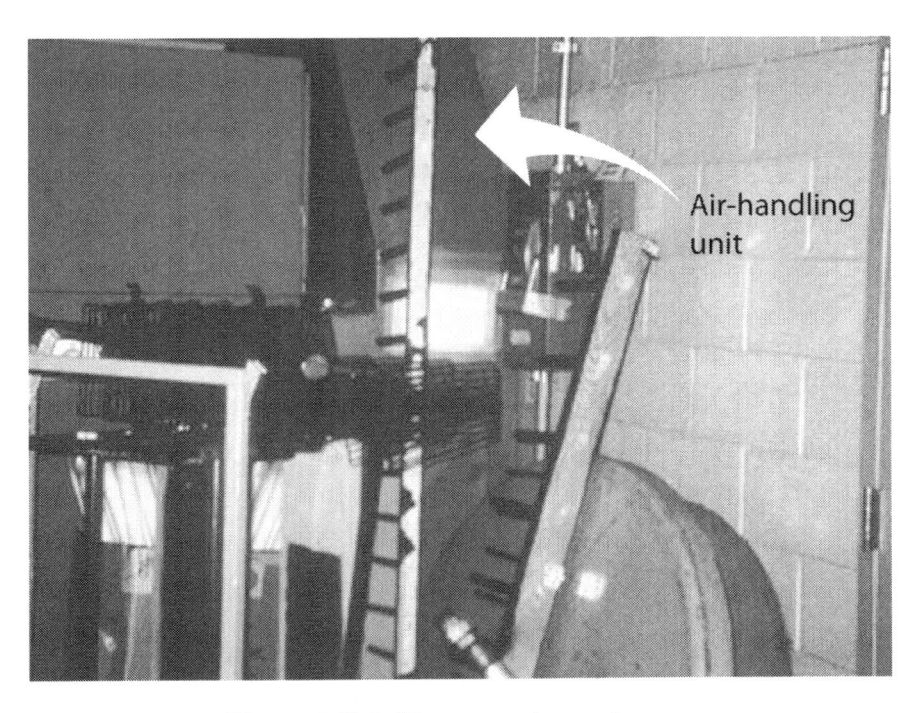

Figure 2.H.1 Messy equipment room

Back behind the pile of junk in the above photo is an air-handling unit. Every time the maintenance personnel need to change filters, grease, check belts, clean the coils, etc. they have to deal with moving all this crap. It can just ruin the day!

In the boiler room, vacuum, or wipe clean everything, then wash down what you can. Add a new coat of light colored paint! Color-code the pipes, if you really want it to look neat. This is your workroom. This is your office. Make it a pleasant place to go to work. I could go on and on about cleaning and reclaiming your boiler room and equipment rooms, but a picture is worth a thousand words. Three pictures are even better.

Figure 2.H.2 Clean, pleasant place in which to work.

Figure 2.H.3 Imagine the lasting impression an equipment room like this makes!

Figure 2.H.4 Work at it long enough, it can even be fun! Wouldn't that be a great place to work?

One final reminder, then we're done with this topic. After the first of November each year, approximately 50% of my company's service calls were dirt related. Air-handling unit and unit ventilator filters need to be changed by this time, every year. Unit ventilator filters need to be changed, at least, four times a year. Air-handling units that operate year-round should have their filters changed six times a year. Combine improved filter maintenance, with keeping your boiler room and equipment rooms clean, and you'll be well on your way to an effective program of providing energy optimization, comfort and indoor air quality, as well as making a good first impression.

Change methods of operations

3.A – Reduce boiler operations

We now start Step 3 of the six-step outline for energy optimization, comfort improvement and indoor air quality. We have discussed, at length, the topics of fixing what's broken and cleaning what's dirty. After years, and even decades, of inadequate budgets, each new generation of maintenance personnel is frequently asked to invent new ways of operating the HVAC system, in an effort to cover up for what's broken and dirty.

So, with decades of no one having operated a building that wasn't broken, or dirty, this usually means that, as facilities fix what's broken and clean what's dirty, the facilities continue to be operated the same way someone deemed they needed to be operated, before the improvement program started. Such practices waste energy, cause discomfort and lead to poor indoor air quality. So, in this chapter, I begin the process of recommending changes on how the facility can be operated, in order to reflect the fact that the HVAC equipment is no longer broken or dirty.

Let's start by talking about reducing the runtime of a boiler. In other words, turning your boilers off, when not really needed. It really can happen more often than what you might think. I'll tell you up front that some of what you're about to read may sound a little overboard. That's okay. Sometimes you have to have to push pretty hard on an obstacle to get it to move just a little. Keep in mind that I'm not a mechanic. This is information about which I have either read elsewhere, or my mechanics have shared with me about boiler runtime reduction. In all cases, consult with the boiler service company, if you have any doubts about something I suggest.

I'll start with noncondensing hot water boilers. The first problem that you want to avoid, when reducing boiler on time, is thermal shock. One way to experience thermal shock is when you have a high fire and cold water in the boiler. These conditions cause rapid expansion and contraction of the tubes, which shorten their life and may cause leaking in firetube applications. Boilers need to be brought up to operating temperatures slowly. With modulating burners, the addition of a high fire lockout control will keep the burner in a low fire mode until the water temp is nearly up to the minimum desired water temperature of 160°F.

If you have a hot water boiler, with a single firing rate burner, you can't just turn the boiler on and walk away. You need to turn it on for about 5 minutes, and then let it sit for 10 minutes, gradually increasing the on time until the boiler is up to temperature. You should plan on this taking upward to several hours.

Most steam boilers have modulating burners. In my opinion, all steam boilers should be equipped with a high fire lockout device. Again, this device would be wired to keep a burner from operating at anything other than the lowest firing rate until the water temperature exceeds 190°F. This helps to avoid thermal shock.

As mentioned earlier, another problem to avoid is condensation. The products of combustion are carbon dioxide and water. If the boiler water temperature is maintained at temperatures above 160°F, you can be comfortable knowing that water will not condense in your boilers. While not as big a problem when burning gas, the sulfur in oil can combine with condensation, forming sulfuric acid that can take years of life out of any boiler.

Now, as I'm certain you already aware, for the majority of the heating season, most buildings can heat quite sufficiently with much less than 160°F water. To maintain the high boiler water temperature necessary to avoid condensation, yet circulate just the reduced temperature water that is needed, the building must have a valve that mixes hot boiler water with cooler return water. Historically, this has been done by using a 3-way mixing valve. But, return water to a fire-tube boiler should never be colder than 30°F less than the water in the boiler. That too is another way to experience thermal shock. To guard against this kind of thermal shock, I recommend a 4-way mixing valve, explained in greater detail, in Chapter 5.C.

As you'll see in the above mentioned discussion, reducing the supply water temperature from the high levels needed for morning boost; to the reduced temperatures needed during occupied times; and even further reduced temperatures during unoccupied times, will significantly reduce the boiler run time. To add to the reduction, consider pump optimization. Again, check the discussion to see how to reduce the operation of heating pumps by up to 1,800 hours a year, in areas with temperature similar to those in Minnesota and Wisconsin. When the pumps shut off, the boiler will run only sparingly to maintain the minimum water temperature, until the pumps come on during early morning boost.

If not using condensing boilers, but you like the idea of matching the boiler water temperature to the actual heating needs, then the use of a 4-way mixing valve to keep the return water from shocking the boiler, is a definite must.

Typically, when an intermittently occupied building is put into night setback, all of the heating valves are driven closed, until the setback temperature of that particular area is achieved. At modest outside temperatures, with all exhaust fans properly turned off, most buildings won't reach the night setback temperature overnight. Most won't even drop to the night setback temperature over a weekend. Sorry if you've heard about turning off exhaust fans when not in use more than you might think is necessary, but it's really important! So, with all the proper minimum water temperature and high fire lockout controls in place, why not turn the steam boiler off for all those hours of operation, as well? It could easily add up to the same 1,800 hours that we see in Minnesota and Wisconsin for pump optimization.

What else can you do to reduce a boiler's run time? Let me relate a story. A good sized senior high school started with two large-box steam boilers. At some later time, with a building addition, another smaller boiler was added. On my first visit to this school, the original two boilers were running, and the third boiler was online, ready to run, if the steam pressure dropped.

With steam blowing out of the condensate return tank, it didn't take much effort to convince the business manager that steam trap repairs, or replacements, were in order. Work started immediately.

During our work, we convinced the maintenance personnel that the steam pressure could be reduced to just 5# and there really wasn't a need for all three boilers to be online at the same time. Even before having an opportunity to eliminate simultaneous cooling and heating in the air-handling units, when the steam traps were repaired, the building operated with only the third and smallest boiler.

With a close relationship with the business manager, we were able to keep track of the change in the natural gas consumption. The savings from just two weeks of operation paid for each day we worked, parts included! That's a pretty nice return on investment (ROI)!!

Now, while we had a great relationship with the business manager, that wasn't exactly the case with the in-house maintenance personnel. Frankly, they were embarrassed about the savings we were able to demonstrate, with such a small investment. Unfortunately, this embarrassment lead to the maintenance staff putting extreme pressure on the business manager to sever our relationship. I wish I could say that this was an unusual circumstance, but it was not. That's why you will frequently read my admonition to let your specialty contractors work with your in-house personnel to build a relationship of know, like and trust. At a later meeting with the business manager, he was quoted to say, "You're not in trouble in our district, because you didn't perform. You're in trouble because you did exactly what you said you would."

So, always ask why the boiler is still on, when there is no need for heat.

I hope that you're seeing the value of the information being provided, but I can't do it all. I'll, again, refer you to another helpful website on boilers at: www.heatinghelp.com. Dan Holohan, a noted author and speaker, developed this site. We were first introduced to Dan through some of our industry publications. I think you'll enjoy some of Dan's insight.

Change methods of operations

3.B - Lower steam pressure

Let's continue with the number one heating energy consuming device, the boiler. At what pressure are your steam boilers operating? Is it more than 5 psi? Why? From my experience, the only thing that higher pressure can do is to cover up problems that a lower pressure can't. As an example, I once called on a school that was operating at 9 psi. When asked why, the maintenance personnel said that they couldn't maintain the proper domestic converter water tank temperature with anything less than 9#.

Looking at the steam supply pipe feeding the converter, I saw that the original paint had never been disturbed. With about 15 years of operation on the system, I suggested the possibility that the heat exchanger tube bundle may be coated with dissolved solids from the fresh water supply. It turned out, in order to remove the tube bundle, through the hole in which it was installed, someone had to crawl inside the tank and knock the chunks of mineral deposits off. After further chemically removing the remaining mineral deposits, the steam pressure was reduced to a maximum of 5 psi, with no difficulty in maintaining proper domestic water temperatures. This was in a 200,000+ square foot facility. Imagine how much could that have been costing them for all those years?

By the way, even with the right answer to their problem, I still hadn't developed the relationship to the point where they knew, like and trusted me. Someone that had been ignoring the issue for years got the work. I eventually got **some** work, but I think we both missed out. Funny how long it sometimes takes!

Excessive steam pressure is often the result when a facility has lots of defective steam traps. The steam short cycles into the condensate drain, before it reaches the end of the supply line. It then requires higher pressure to get steam to the far reaches of the building. After replacing defective steam traps, with the recommended long-life TLV steam traps, the steam pressure can then be lowered to a maximum of 5#.

I know that you've all heard about thinking outside the box. Well, here is another one to think about. Over 30 years ago, I was working with a multi-building complex, using a central steam plant, operating at 125#. At each remote building, the steam was reduced to just 5#. When asked why, everyone said, why not? It's been run that way for decades. Why should we do anything any differently now?

I suggested that, perhaps, the remote buildings weren't so remote from the central boiler; that the steam pressure could be significantly reduced; and there would still be 5# steam at all of the remote facilities. When the central plant lowered the pressure to 50# steam, they did have to replace a couple of end-of-main steam traps, but the cost of operation dropped like the proverbial anvil in a swamp. Everyone thought that they were heroes. I can only wonder how many more decades it will be before someone tries dropping the steam pressure to 25#, then perhaps 10#, or maybe even less, without the pressure at each facility not dropping below 5#. I encourage you to question everything! You may be surprised as to how well it turns out.

How are you operating your steam boilers? Do you have more than one boiler on line, at all outside temperatures? Why? We encourage you to look at nearly every operating procedure and ask why? "That's the way we've always done it.", is not a proper answer. With the tremendous internal heat gain available in most occupied facilities, especially schools, space temperatures will not drop rapidly with the loss of a boiler. When the outside air temperature is above 32°F, the indoor temperature of very few facilities will drop to 65°F overnight, or even over an unoccupied weekend. Once you prove that to the in-house maintenance personnel, it becomes easier to implement changes in the immediate availability of a backup boiler. In summary, always ask why. Keep asking until you get an answer that makes sense.

Changes in operations

3.C Hot water reset

Hot water heat can come from multiple different sources; steam to hot water converters, district heating converters, or hot water boilers. All should be capable of changing the supply water temperature as the outside temperature changes. Just remember, the converter temps and condensing boiler temps can be reset directly. For non-condensing boilers, you're looking to change the reset schedule on the control for a 3-way mixing valve.

The typical hot water reset controller, at least in the upper Midwest, is set to maintain approximately 100°F water temperature, at 60°F outside air temperature, and moves linearly to approximately 200°F water temperature, at −20°F outside air temperature. Under this type of reset schedule, rarely does anyone ever have cold complaints, due to a lack of hot water. Because this system is rarely an area of complaint, it is seldom looked at as an item for more efficient operation.

In chapter 5.B, I'll spend a lot more time on this topic, but for now, find out what your reset schedule is. If the -20°F outside air temperature reset water temperature is more than 160°F, or the 60°F outside air temperature reset temperature is more than 100°F, start gradually lowering the reset schedule. When comfort complaints arise, investigate the specific area to determine the source of the issue. Is something broken, dirty, or is the system providing simultaneous heating and cooling? Solve the issue and keep lowering the reset schedule.

Change methods of operations

3.D - Night setback

I frequently find that night setback controls are no longer functional. Sometimes the systems are broken, but more often, they have been deliberately abandoned. Properly operating night setback control saves approximately 15% to 22% of the annual cost of heating, and even more on cooling, so it's important that occupied/unoccupied controls be kept in good working order. This chapter will help explain why we find these conditions. I'll also explain the mathematics behind how the savings are achieved.

Many people already know, and understand, the importance of night setback, but they just don't know that their system is broken. If you have a building automation system, BAS, you should be able to check the space temperature trend log of various locations. If you do not have a BAS, a valuable, yet inexpensive, tool is a simple temperature chart recorder. Simply place the recorder on the top of a desk, or a shelf, for a few days. If the recorder is indicating no significant change in space temperature when the facility is unoccupied, you will have good economic justification in finding out why and taking corrective action.

The problem that leads to most night setback systems being abandoned is the inability to recover from a reduced night setback condition. 99% of the time, that's caused by not fixing what's broken and not cleaning what's dirty. We are well into this training program. If recovery from night setback is still difficult, you may want to revisit steam trap repairs, coil and blower cleaning, belt conditions and adjustment, and so on. Trust me - it's not rocket science!

Now, here is the mathematics behind night setback. It's all based on the basic formula for calculating heat loss:

Btu's of heat loss = square footage of heat loss area X U X Delta T (ΔT)

U = 1/R, R being the insulation rating of the walls, windows and roof

ΔT = the difference between the inside and outside temperatures

In this formula, two of the three factors are not readily changed from occupied to unoccupied times. Only the temperature difference can be easily changed. In Minnesota and Wisconsin, the average outside temperature, during the heating season, is approximately 28°F. When we maintain an indoor temperature of 68°F, it gives us a ΔT of 40°F. If 100% of the heat loss is affected by a 40°F ΔT, that means that each degree that you drop the space temperature affects the heat loss by 2.5% (100% divided by 40°F).

When you drop the space temperature by 10°F, it equates to a 25% reduction in heat loss during those hours. It's simple mathematics.

I was trying to explain this without repeating something you will see later, in Chapter 5.C, but it's really needed here. Since you will be seeing it again, I will try to abbreviate it here.

Referring now to the desired building temperature profile, for an intermittently occupied building, in Figure 3.D.1, line AB represents an occupied space temperature. Line CD represents the time during which the night setback temperature is maintained. On a continuum, points F & A are the same.

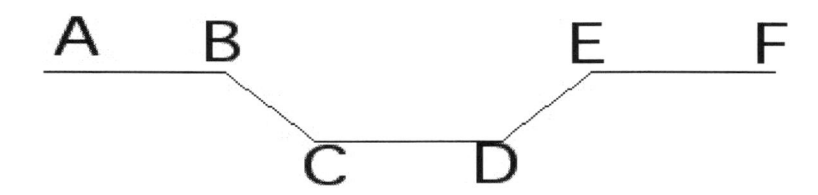

```
A      B                    E      F
_____\                  /_____
         \                /
          _____/
          C              D
```

Figure 3.D.1 Profile of the desired day/night temperature profile, for an intermittently occupied facility.

The most frequently heard argument against night setback is, the energy I save going into night setback, Line BC, is the exact same extra energy that I have to put back into the building during early morning recovery, Line DE. As far as that statement goes, they are absolutely correct. What needs to occur is to maintain a reasonable night setback temperature, Line CD, for an extended period of time. If the heat source is simply turned off every night, and turned back on in the morning, there will be no savings from the drop in space temperature. That is, Points C and D come to a point at the bottom of the two slopped lines. You might save some electrical energy by not running the fan, but there won't be any savings from night setback!

Having said that, one must also consider the fact that air-handling systems also provide ventilation. When the air-handling units are in a proper night setback mode, no ventilation is provided until the time of occupancy, which can save huge, additional amounts of energy. The combination of reduced space temperature; the elimination of ventilation during unoccupied times; coupled with the varying length of unoccupied times, usually works out to 15% to 22% reduced heating energy cost.

So, how many degrees should you set back the temperature? Well, here is where art is combined with the science. By using a graphic temperature log, either by BAS, or chart recorder, the first thing to look at is the rate at which the temperature drops. When the outside air temperature is above 35°F, most commercial office buildings, and schools, won't drop the usual 10°F setback overnight, or even over a weekend. If you find that the temperature is dropping faster than that, check the operation of all the exhaust fans. Something is prematurely pulling the heat out of the building and it's most likely that some exhaust fans aren't being turned off at night.

When it is cold enough to reach a 10°F setback, look for a dogleg in the temperature profile graph. The temperature of the air will drop at a pretty steady rate for several degrees. At some point, however, the rate of temperature drop changes, distinctively, to a slower rate. It is at this point that the air starts taking heat from the walls, floor, ceiling and furnishings. My experience is that this is the temperature that should be used for a target, heating season, night setback. If you go to a lower setback temperature, it is too difficult to return the heat to the furniture, by the time of occupancy. People will radiate heat to these cold surfaces and comfort complaints will abound. For most installations, I find the temperature, at which this dogleg occurs, is usually about 6°F lower than the occupied temperature.

So, if you have a pneumatic temperature control system, or a simple electric control system, with just one setback setting, use the temperature at which the dogleg occurs. If you have a building automation system (BAS), your system should be capable of a flexible night setback operation. Use a 10°F night setback setting on Friday evening through sometime very early Monday morning. If your BAS has the ability to optimize that morning boost time, great. If not, just add a couple of hours to your Monday recovery time. Your occupants will let you know if you didn't add enough time. Again, use the dogleg setback temperature on Monday through Thursday.

Okay, what about air-conditioning season setbacks? Well, you'll never hear anyone complain about early morning furniture being too warm. All you usually need to do is have cool air blowing on the occupants when they arrive. To be on the safe side, start the air-handling units and A/C about 30 minutes before the time of occupancy. Just be certain to delay the start of ventilation until the time of occupancy.

If you have a BAS, you can start your air-handling systems early, like midnight kind of early, if the outside air temperature is at least a couple of degrees cooler than the inside air temperature and the outside air enthalpy is less than the inside air. More on enthalpy control in Chapter 4.B.

For your home setbacks, I recommend just a 4°F setback for both unoccupied times during the day and night time sleeping hours. In my opinion, the time duration of a home setback just doesn't warrant greater setback temperatures.

Change methods of operations

3.E - Lighting

Now, let's do something in memory of dear old mom and dad. They were always telling us to turn the lights off, when we left the room. So, how long do you have to leave lights off to warrant turning them off? Well, I'm certain that with the new lighting systems being introduced almost daily, the rules change constantly. But, what has always been said in the school industry was, "If you're going to the bathroom, leave the lights on. If you're taking reading material with you, turn the lights off!" Start there until you have an opportunity to check with your favorite lighting specialist on your specific lighting systems.

Many facilities have someone that is in charge of rewarding occupants for turning off their lights, when not needed. Initially, they will leave a small note, and a piece of candy, thanking them for their contribution to helping control energy costs. Later, they just leave the candy. People will know why they got it.

Figure 3.E.1 Security and decorative lighting are more great places to use time clocks and building automation system outputs.

Figure 3.E.2 Light sensing time clocks can help make life a lot easier too.

Just like I keep telling my 14-year-old grandson, you can be hugely successful in this world, if you just learn to pay attention. How many times do you suppose the maintenance staff have walked through this parking lot, without noticing that the lights are on during a cloudy day?

It's an old technique, but you can also just take out some of the bulbs. Back in the mid 70's, one facility, with which I was familiar, simply removed the inside two tubes on all 4-bulb fluorescent fixtures. A total of over 14,000 bulbs were removed, just from that one building. Bulbs were removed after normal working hours and no one ever mentioned the change in lighting levels. Always check with your lighting specialist to make certain that what you're doing is okay.

As electricity gets more expensive, fluorescent bulbs get more efficient. Any time you get to reduce wattage, without adversely affecting lighting levels, give it a try.

Another quick energy saving procedure is to simply reduce the number of fixtures, in areas that have too many. I consistently see hallways that are over lit.

Change methods of operations

3.F - Eliminate Electric Boost Heaters

In larger commercial facilities, the cost of electric energy is based on how much you use, as well as how fast you use it. Electric companies monitor the amount of electricity you consume during a 15 minute sliding window. Each month, commercial accounts receive a demand charge, based on the highest 15 minute period of consumption.

The electric companies have to be ready to meet your demand, any time you want it. Therefore, the demand charge is based on the highest 15 minute period for the month, or something like 85% of the peak demand during the previous 11 months, whichever is higher. This high demand peak almost always occurs during the hours of 11:00 A.M. to 2:00 P.M., when the kitchen is in peak operation. So, if someone wants to test the lights on the football field, or in the parking ramp/lot during this time, you may be paying for that mistake for the next 11 months!

One of the things that I've never been able to figure out is the process of selecting electric kitchen appliances for schools and commercial kitchens. This includes electric boost heaters for dishwashers. Then, shortly after the building is occupied, they try to figure out how to eliminate them, because the first electric bill, with high demand charges, arrived.

In new construction, I encourage you to make certain that life cycle cost comparisons are always made in the selection of kitchen appliances. And, be certain to include the comparison between a separate gas water heater for the dishwasher rinse cycle, or even chemical sanitizing, and electric boost heaters. In retrofit situations, we can most frequently find locations for the installation of gas water heaters, someplace close to the kitchen. One would think that the same could be done in new construction too.

Fortunately, I don't see it much anymore, but it used to be pretty common to see a separate steam heated, dishwasher booster coil in the boiler room. Most had their own circulating pump. And, most of those pumps were left on 24/7, wasting a lot of energy. In addition, this often lead to turning the boiler(s) on long before needed for heating. All domestic hot water circulating pumps need to be turned off when the facility is unoccupied. Dishwasher pumps can usually be turned off even sooner.

Avoiding the usual peak electric demand times, by limiting the use of other electric devices should be reviewed too. One of the largest, and most frequently abused of these other devices, is the kiln, used by the art department in many schools. The typical scenario is for the art teacher to turn the kiln off upon his or her arrival and open it to let the contents cool. Things are usually cool by 11:00 A.M., reloaded and turned on, just in time to add to the peak electric demand caused by the kitchen equipment. We encourage you to discuss the operation of the kilns, as well as any other high electric demand equipment, with the people in charge. They need to change their times of operation to periods of reduced electric demand. If that doesn't work, their use can be controlled by time clocks, building automation or padlocks!

Change methods of operations

3.G - Summer water heaters and swimming pool heaters

This chapter primarily relates to many older schools, but commercial building owners and managers may still find some nuggets of information for their facilities, as well.

After all the energy conservation work that has been done in so many schools, I'm surprised to still find heating system boilers still running during the summer months. Standby losses from oversized boilers can be huge. I really encourage you to operate these systems as efficiently as possible, or consider replacing them altogether.

Let's start our discussion with elementary schools. Many still maintain large, domestic hot water storage tanks, heated by steam to hot water converters. The purpose of a storage tank is to maintain a reserve of hotter than necessary water, so that the minimum water temperature needed during peak times, is still maintained. To date, I've never seen an elementary school that has ever needed such a system. Elementary age students don't take showers. Standard, 80 gallon, gas water heaters have always been found to be sufficient for serving the needs of elementary school kitchens and lavatories. Later in this chapter, I'll have more on how to serve dishwashers and lavatories, from a too hot water heater or storage tank that may be necessary for peak consumption times. For now, just know that you should be able to completely remove the large storage tanks, in most elementary schools.

When it comes to large storage tanks in high schools, they too are almost never needed. Kids just don't take showers like they used to. I don't know how they ever get dates. You simply need to monitor the storage tank water temperature to see if they are ever needed. If the temperature of the tank doesn't change all day, then heating the storage tank is a huge waste of energy.

The two most frequent situations, for which we find high schools using heating system boilers in the summer are, swimming pools and facilities that have, or believe that they have, a need for heating with the HVAC system. Let's start with the swimming pool.

Steam to hot water swimming pool heaters are supposed to have valves that automatically close, if the pneumatic or electronic control signal is lost. You don't ever want anyone jumping into a too hot swimming pool. If you closely monitor these control signals, you will find that the control valve is rarely, if ever, very far open. That is, the steady-state heat loss from a pool is usually very low, in comparison to the control valve's capability to heat the huge amount of fresh pool water rather quickly. Therefore, we suggest that, if there is no separate pool water heater, the building heating system boiler be turned off for a couple of days between operations. My experience has been that operating a steam boiler for four hours, every second or third day, can maintain pool water temperatures within 1°F of setpoint. Of course, low boiler water temperature control and high-fire lockout precautions, as discussed previously, apply.

The same type of operation also applies to the large, domestic hot water storage tanks. Monitor the minimum temperature of the storage tank. If the temperature is always above the minimum, turn the boiler off until heat is actually needed. Again, we have seen times, in the summer, when running a boiler for 4 hours, every third day, has been sufficient.

Again, once the pool is heated, converter control valves operate in a mostly closed position. As discussed previously, this leads to wire draw on the valve plugs and seats. I suggest the installation of a manual bypass valve, in parallel, with a small, steady-state heat loss sized valve needed for the pool. Once every few years, when the pool has been drained and refilled, you simply open the manual bypass valve, to add extra heat, until the water temperature gets close to the desired 80°F setpoint.

In systems where they have both a swimming pool and a large domestic hot water storage tank, one or the other will require heat first. You'll soon learn to monitor that one most closely and run the boiler accordingly.

It's been years since I've heard anything of the deadly Legionella Bacteria, originally discovered in an air-conditioning cooling tower. However, according to the Minnesota Health Department, the same bacteria have been found in large, domestic, hot water storage tanks. Prior to the energy crisis, most domestic storage tanks were kept at temperatures near 140°F. Maintaining large tank temperatures at, or above, 135°F keeps the bacteria in check. Turning the temperature of these tanks down to the 105°F range was a quick and easy way to save energy. And, while I have not personally heard of any Legionella Bacteria problems caused by this, the problem is still an apparent potential.

When storage tank water temperatures are kept low, there is no concern of scalding. Therefore, for the past several years, with no news of Legionella Bacteria, few provisions have been made for returning the temperature of domestic hot water storage tanks to one that kills the potentially hazardous Legionella Bacteria. However, simply turning the tank temperature up past 135°F, to kill the bacteria, is not sufficient to address the additional problems of energy consumption and scalding.

When hot water usage is sufficiently large enough to warrant the use of large storage tanks, the solution is to install a 3-way mixing valve that mixes a portion of the potentially scalding hot water, with some cold water. The resulting water is then maintained in the safe temperature zone of 105°F to 110°F.

When it comes to providing enough hot water for the rinse cycle of dishwashers, electric boost heaters are often the low cost, first choice. However, as soon as the first electric bill shows up, with a significant increase in the demand charge, a more expensive, initial first cost gas water heater, solution is often investigated. I encourage facilities to fully investigate all solutions prior to new construction, as well as to investigate life-cycle cost-effective alternatives for existing facilities.

Oh yes, don't forget to add the ability to turn off the hot domestic water circulating pumps, during unoccupied times. Since this is a regular occurring item, a time clock or a digital output from the automation system is in order.

I also made mention of the summertime use of boilers for HVAC. I will go into the elimination of simultaneous cooling and heating in great detail, in Chapter 4. At this point, I will simply ask if you have ever tried to control the air-conditioning temperature in your own home by opening more windows? Summertime use of boilers for your HVAC system makes about as much sense. It can be 100% eliminated in all office and school situations.

Before leaving Step 3, I want to emphasize that the list of items that I have provided to monitor, and potentially change, is only a partial list. This list is intended to be thought provoking, not necessarily a complete checklist.

Change methods of operations

3.H - Install timers and time clocks

When budgets do not allow heating, ventilating, air-conditioning and temperature control equipment to be fixed when broken or cleaned when dirty, it causes the building engineer to operate the system in whatever manner is necessary to try to provide some semblance of proper comfort. One of the most frequently abused items to do this is time clocks.

In order to cover up for the fact that the coils are dirty, steam traps are in ill repair, filters are plugged, etc., many systems need to start earlier in the day to achieve comfort levels at the time of occupancy. Therefore, we frequently find start times set much earlier than what was originally deemed necessary. In addition, stop times are also frequently, and unnecessarily, changed to extend the daytime operation.

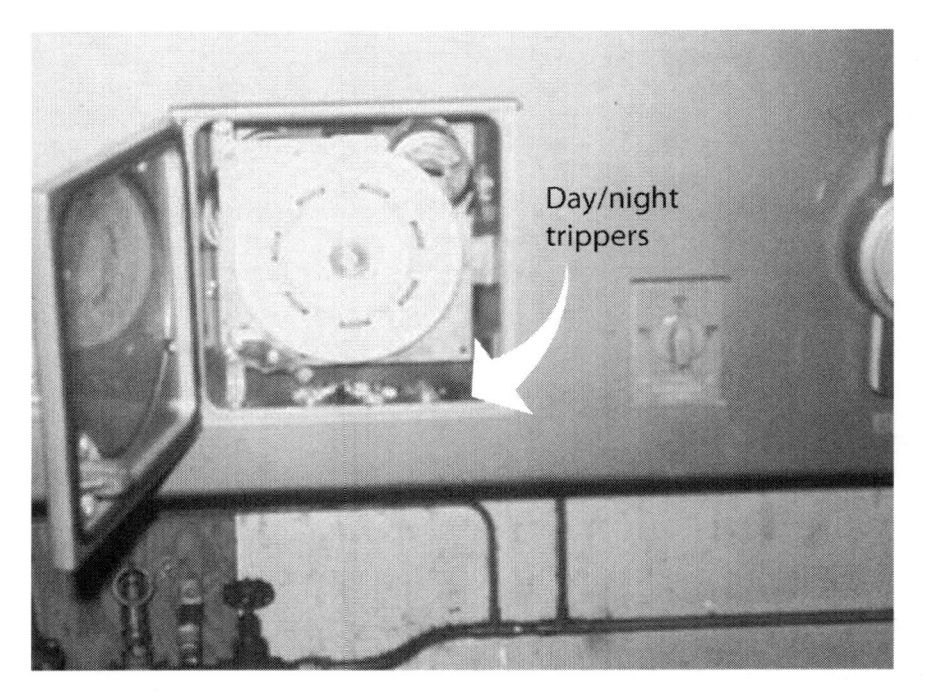

Figure 3.H Day/night time clock with all the trippers disconnected.

On more occasions than anyone would like to admit, I find all the day and night trippers lying in the bottom of time clocks. This allows the associated equipment to operate 24 hours a day, 7 days a week. Find out what caused someone to do this and take corrective action. Here's a hint - Look for something broken or dirty!

Whenever a power outage occurs, or daylight saving time changes, all mechanical time clocks need to be reset. Not having the correct time on the time clocks frequently leads to early morning or late afternoon comfort complaints, for which there is no obvious reason. Once the correct time is restored, everything returns to normal. Right after daylight saving time changes, we encourage everyone to check and correct all mechanical time clock settings.

By the way, do you know where all your time clocks are located? It would probably be a good idea to have them all written down on a list, so that others can find them. In other words, prepare for your long, enjoyable and uninterrupted retirement, well before it happens.

There are many situations where, what you're looking to turn off is only operated intermittently. Welding hood exhaust fans, or short-term chemical experiments, come to mind. In these situations, the addition of a simple spring wound timer may suffice. They come in a variety of on-time durations and are as easy to install as replacing a light switch.

The bottom line is that all of these situations indicate an operation that may be covering up for system deficiencies. It's costing you a lot of money! Make a list of all equipment you have and establish occupied hours. Right after daylight saving time changes is also a good time to tour your facilities. Make one tour late at night and another early in the morning. Make a list of what's running. Make a determination if it should be running. If the present operation is incorrect, add timers, time clocks, or building automation.

Whether you're dealing with school budgets, or commercial building budgets, make a list of your needs. Next, since you'll never get all the money at once, you'll need to prioritize the list. How do you prioritize? Simple, just follow the Six-Step Maintenance Recovery Program that we've been working on:

1. Fix what's broken

2. Clean what's dirty

3. Change methods of operations

4. Revise temperature control sequences

5. Install new technology hardware

6. Implement preventive maintenance routines

From the parking lot to the roof, the basics of this outline can be used for nearly every part of your facilities.

Change methods of operations

3.I - Ultrasonic fault detector

In Chapter 1.B, I talked about the economics of steam trap testing and replacement of defective steam traps. Remember, at $0.70/Therm for natural gas, you'll save about $210 a year by replacing a defective steam trap, in an application where there's a valve on the heating coil. In just about every situation, even without a utility rebate, that's less than a one year return on the investment. You'll save about $1,050 a year for every defective end-of-main or header trap you replace. The return on that investment should be measured in months, not years. So, if you have a limited budget, guess which traps need to be done first and soon? The replacement of those traps will quickly fund the replacement of all the others!

Testing steam traps is generally done with the use of an ultrasonic fault detector. Steam flowing through a restriction, such as a steam trap, makes an ultrasonic noise. An ultrasonic fault detector works like an electronic stethoscope, with a speaker attached.

Figure 3.I.1 Son-Tector ultrasonic detector 123 package

With the ultrasonic unit on, you can place the contact probe on your shirt and drag it across the fabric. The noise you hear is similar to what a bad steam trap sounds like.

When testing steam traps, be certain the steam trap is exposed to steam. With end-of-main traps, all you need to do is to make certain that the boiler is on and that the trap is hot. Sometimes the strainer ahead of the trap is plugged, and if you don't check to see that the trap is hot, you may think a quiet trap means it's okay, when you really don't know what its condition is. To test the trap, you simply put the contact probe on the discharge side of the trap.

With traps on heating coils, it's a little more involved. If you're testing traps on fan system coils, turn the fan off and open the heating valve. While you're waiting for the steam to reach the end of the coil, you can start learning about testing, by just listening. You'll soon get used to hearing what things sound like. After you've gained some experience, you might want to leapfrog ahead to turn the next unit off and open the valve, before returning to the previous unit.

It's the same with converters. That is, you need to have steam in contact with the steam trap in order to test it. Turn the pump(s) off and open the steam valve. Be certain to test domestic hot water converters after most of the people have gone home. You don't want to scald anyone.

After testing the traps, drive the heating control valve closed. Visually inspect the valve for steam leakage around the valve stem, as well as other physical damage. Then test the control valve for steam leakage, by placing the ultrasonic contact probe on the body of the closed valve. Again, a control valve with a damaged seat, and/or plug, will sound the same as a non-functional steam trap. Document the make, model number and pipe size of any defective items for future scheduled work.

Most ultrasonic fault detectors have both a contact probe, as well as an air sensor for detecting air leaks. Before leaving the area, put the air sensor on and check for a leaky diaphragm on the heating control valve. If there is any air pressure on the damper motors, check those for air leaks too. Then, wave the air sensor around the room to see if you can detect any leaks in airlines going through the room. Also be certain to open all control panels and check for air leaks.

What's an air leak sound like? Try saying "soup, sandwich and other similar sounds" into the air sensor. The ultrasonic noise will be very similar to the noise made from an air leak in the pneumatic temperature control system. You can even have someone at a considerable distance make that sound. It's pretty amazing how far away you can hear a small air leak.

My servicemen have found leaky classroom and office thermostat mountings while walking down the halls, with the ultrasonic unit strapped to their belts. They have found air leaks above drop ceilings, just by having the unit on, pointed towards the ceiling, when walking from one equipment room to another.

One school district, that had their own ultrasonic fault detector, said that they found copper air lines that were not properly soldered together, more than 20 years earlier. Another school district took a temperature control air compressor from a 75% runtime toa 25% runtime, just by repairing all the air leaks. The electric savings alone paid for their ultrasonic fault detector in less than one year!

In addition to listening to what's going on inside steam traps, the contact probe of an ultrasonic fault detector has also been a great asset in listening to bearings. Dry bearings make a fair amount of ultrasonic noise. And, damaged bearings, even well packed in grease, still usually make enough ultrasonic noise to distinguish them from good bearings.

Testing bearings is one of the least used features for the ultrasonic fault detector. However, it has the potential for saving the most amount of work. I have actually seen a 4-inch diameter shaft, 8 feet long, worn to the point where the air-handling unit fan actually started rubbing against the housing, before any problem with the bearing was noticed.

For openers, how many mechanical rooms do you have where an 8-foot shaft can be removed without punching a hole in a wall, or worse, having to move equipment from the second room? Then the shaft had to be sent to a machine shop, where it was rebuilt and machined back to the 4" diameter, before it could be reinstalled, along with a new bearing. In this case, an ounce of prevention certainly could have saved a ton of work!

Another memorable bearing application was a school that had a problem with a boiler burner. They examined it closely, with their own ultrasonic fault detector. Finding a faulty bearing, they were able to call the manufacturer with the make, model and precise description of the bearing location. A few days later, with parts in hand, the school's own mechanic quickly replaced the defective bearing.

That success story led this school district to survey all the air-handling unit bearings in the district. We never heard the final tally, but at one time, they had six major air-handling unit bearings that they had scheduled for summer replacement. This is a classic example of labor productivity being doubled by having a program of planned preventive maintenance, with scheduled parts repair and replacement. If left undetected, guess when the replacement of these bearings would have been discovered. Right, the most inopportune time imaginable!

Now, my company maybe missed a lot of additional opportunities by not using our ultrasonic fault detectors to find electrical problems, but you don't have to. The website for the fault detectors we used, listed at the end of this installment, has numerous articles to help expand ways for you to use your ultrasonic fault detector.

Do you have one, or more, leaky tubes in a fire tube boiler, but don't know which tube? Open the doors to the tubes on each end of the boiler. Then, put an ultrasonic generator in one of the cleanout holes, near the bottom of the boiler and put the cover back on the hole. Using the air sensing probe, check each tube, and the rolled ends of the tubes to hear from where the ultrasonic noise is coming.

The more you use this device, the more applications you'll find for its use.

Bottom line? Ultrasonic fault detectors can be a valuable tool to every large facility, or complex of facilities. It is an absolute essential to anyone working on pneumatic temperature control systems.

For information on owning your own ultrasonic fault detector, and how to use it, contact:

Ansonics, Inc.
1307 Paseo del Pueblo Norte
El Prado, NM 87529
www.ansonics.com
575-758-4555
sales@ansonics.com

Revise temperature control sequences

4.A - Eliminate Simultaneous Cooling and Heating

As the name implies, the purpose of heating, ventilating and air-conditioning air-handling units is to heat, ventilate and cool. The key is not doing all three at the same time! Ventilating during occupied times is a constant. Then, using more outside air for atmospheric cooling is the low-cost way to add comfort, prior to adding a variable amount of air-conditioning. When heating is called for, it's important to make certain that, to the best of the system's capability to keep everyone comfortable, no cooling, especially costly air-conditioning, is being provided.

So far, the first three steps have all been low or no-cost items. When it comes to temperature control system revisions, it is especially important to consider the big picture of all six steps. Installations of BAS will be discussed further in Chapter 5.D. But, with the rapid explosion, and advancements, in modern, electronic BAS controls, one must question the wisdom of repairing and revising a broken down, dirty, oil contaminated, pneumatic temperature control system. I hate to say it, because pneumatic temperature control systems have done so much for my career, but the fact of the matter is, they are living on borrowed time. No one is installing complete pneumatic control systems in new construction. Pneumatic service mechanics that know how to repair, calibrate and update the wide variety of pneumatic control devices are retiring at a rapid rate. With rare exceptions, they are not being replaced.

So, begin by asking yourself, if I spend significant amounts of money on upgrading the sequence of operation of an antiquated pneumatic temperature control system, is it truly a wise use of limited resources? Is the life expectancy of your building more than 15 years? Is your facility still heated with steam? Is the basic heating and ventilating system capable of meeting today's ventilation codes? Does this facility have acceptable handicapped access? These are all questions that should be asked, before spending lots of money on any phase of recovery, beyond step 3. Now, I want to remind you that there are buildings throughout the world, that are 200+ years old and in the prime of their life! Don't give up on a really nice building, just because it's old.

If major upgrades are in order, now is the time to consider the advantages of a BAS, hot water heat, and improved ventilation. Now, I'm not saying that old pneumatic controls are bad. In the right hands, pneumatic controls can provide nearly identical sequences of operation as a BAS. Fortunately, pneumatic temperature control system revisions can be pretty inexpensive. Done correctly, they can provide a return on investment that can be incredible.

What pneumatic controls can't do for you is to provide trend logs. This is where you select a variable temperature, percentage opening, speed, etc. and the period of time which you want to examine. In seconds, you can look at that trend. Especially early in the life of a BAS, this should be done frequently. Soon after starting in this industry, the smartest control mechanic I ever knew said, "Sometimes the best thing to do with your toolbox is to sit on it." With pneumatics, learning the trend of these values can take hours. With a BAS, these trends are instantly available. Without periodically reviewing your trend logs, there are lots of things that can be going on that you don't even notice.

As an example, a facility, with which I was recently working, had the percentage of outside air bouncing between the minimum setting of 10% and 100%, as many as 22 times a day. At the same time, the variable speed of the supply fan cycled from 50% to 100% and many as 26 times a day. Because this was just one new air-handling unit, serving a large area with three other similar sized air-handling units, the fluctuations went unnoticed! Imagine the wasted energy!!

The first of the following HVAC designs are constant volume air-handling systems, which are still in operation by the tens of thousands, so I'm making the assumption that most are still using pneumatic temperature controls.

Reheat air-handling systems

Probably the oldest design of air-handling units is the reheat system. The air-handling units themselves most frequently have no heating, but will sometimes have a preheat coil, if the mixed air temperature is allowed to go below 55°F. With some thoughtfulness about the elimination of simultaneous cooling and heating, most preheat coils can be eliminated. As the name implies, heat is provided by remote reheat coils, each controlled by a space thermostat.

The air-handling units just provide ventilation, atmospheric cooling and air-conditioning, usually all at the same constant temperature, 55°F, year-round. As you might imagine, a constant supply of 55°F air can be quite uncomfortable. On these systems, all of the space thermostats are remote to the air-handling unit. That means that the least expensive fix is to monitor the return air temperature. It will give you a good indication as to when the space temperature is too warm. Knowing that everything is adjustable; ventilation still needs to be considered; as well as enthalpy control, the controls should establish a reset schedule for cooling that looks something like:

Return Air Temp	Mixed Air Temp	A/C Discharge Temp
60°F	70°F	75°F
68°F	70°F	72°F
75°F	55°F	60°F

Controlling off return air is also helpful in eliminating short cycling of direct expansion air-conditioning too.

Multi-zone air handling systems

A variation on that design is a multi-zone air-handling unit. This system provides cold air in one horizontal section of the air-handling unit and hot air in another. A space thermostat, connected to a set of hot deck/cold deck dampers, then mixes a portion of cold air and warm air, at the air-handling unit, and delivers it through a dedicated duct to the appropriate zone (space), as needed to satisfy the comfort of that area's occupants. One air-handling unit may serve as many as 8 to 10, maybe even more, individual spaces, each needing a slightly different mixture of cold air and hot air.

As originally designed, the cold air is cold enough to satisfy the space condition on the hottest day and the hot air is hot enough to satisfy the space condition on the coldest day. As you might well imagine, providing both, every day of the year, was at a considerable cost of energy. But again, as long as people were comfortable, and the cost of fuel oil and natural gas were in the area of 11 cents a gallon, nobody paid a lot of attention to the inefficiency.

In this case, every thermostat signal is right at the air-handling unit. Your control contractor can easily determine the zone that is calling for the most heating, using it to control the temperature of the hot deck to the minimum temperature needed. They can also determine the zone that is calling for the most cooling, using the hottest zone to sequence the use of atmospheric cooling, beyond the ventilation minimum, and air-conditioning to provide occupant comfort, with the minimum use of cooling energy. This system has the potential to still have some simultaneous cooling and heating, but it will be minimal.

The one thing that may cause discomfort is setting all the thermostats at 68°F. Pneumatic thermostats only know setpoint, plus or minus 2°F. So, this system will be trying to provide full cooling by the time the warmest space temperature is 70°F. In this case, return air control of atmospheric cooling and air-conditioning, similar to the reheat system may be called for.

Double-duct air-handling systems

This is a very memorable story of eliminating simultaneous cooling and heating. An 11 story medical office building had just two air-handling units. Each unit was designed as a double-duct system. One duct carried warm air, at a constant temperature, warm enough to heat any space, at any outside air temperature – commonly 110°F. The other duct carried cool air, cold enough to cool any space, at any outside air temperature – commonly 55°F. In the space, a thermostat selected a portion of each air supply to satisfy the comfort levels of that area's occupants. Metered steam and chilled water, from the hospital across the street, provide the heating and cooling energy.

For each air-handling unit, using existing pneumatic controls, we simply revised the sequence of operation for the heating and cooling duct temperatures, as follows:

At 60°F return air temperature, the heating duct is controlled at 110°F. That's the same temperature that was previously maintained year-round. At 70°F return air temperature, the heating duct is controlled at 70°F. That is, the heating is off. The heating duct temperature modulates linearly between the two setpoints.

At 65°F return air temperature, the mixed air, the combination of outside air and return air, is sequenced above the minimum required for ventilation, with the chilled water valve, to maintain the cooling duct at 65°F. At 75°F return air temperature, the mixed air and chilled water valve maintain the cooling duct at 55°F. That's the same temperature that was previously maintained year-round. The cooling duct temperature modulates linearly between the two setpoints.

Yes, I am aware that not all simultaneous cooling and heating were eliminated. This was a medical office building, with more than its fair share of naked bodies, so we worked hard to help keep everyone comfortable. All settings are adjustable, if needed.

The work was done in just two days. The bill came to $1,100. A week later, I received a call from the person that authorized the work. He asked if I could rebill him. He was only authorized to spend $1,000 on any one item. One bill had to be for slightly less than $1,000 in labor. The remaining bill could be for parts, even though we just reused the controls that were already there. When I hand-delivered the second bills, less than two weeks after the work was done, wewent to the hospital to check on the change in steam and chilled water consumption. The energy savings, in just those few days, had already exceeded the costs. How would you like to make that kind of investment every day? That kind of savings can help finance future BAS work and more.

So, that's a great example of simultaneous cooling and heating. Now let's better define it. It's based on the old theory that, if you blow enough 55F air into any space, you can cool it, even at the hottest outside air temperature. Then, if blowing that cold air into the space makes the occupants uncomfortably cool, as it probably would at outside air temperatures less than the hottest days, you just add a little heat to that cold air. In terms of precise temperature and humidity control, this system is the cat's meow! And, when fuel oil was 11 cents a gallon, its lack of efficiency was of little concern. This system design is still prevalent in many older existing facilities.

Variable air volume (VAV) air-handling systems

Before discussing this air-handling system, I just want to remind you that every commercial air-handling unit is required to provide heating, ventilating and cooling. This system, in its original design, is the first attempt at not providing simultaneous heating and cooling, but it fails miserably at everything else it's supposed to do, especially providing comfort!

From my experience, this is the air-handling systems with the highest levels of comfort and indoor air quality (IAQ) complaints. Keep in mind, I'm old, so when I go back to the original concept of VAV, I'll ask that you give me a little slack. The early VAV systems, with which I am familiar, had a single temperature thermostat that sequenced the flow of hot water through perimeter radiation, with a VAV box that modulated the flow of 55°F air.

At all times, the system brings in enough outside air to meet ventilation requirements, even when the space is trying to heat. For that reason, the VAV boxes are set to a minimum airflow value, intended to provide ventilation, which often makes the space uncomfortably cold. Again, the 55°F air supply temperature is year-round. During much of the year, when minimal cooling is desired, it doesn't take much of that cold air to satisfy the thermostat. Under these circumstances, the most frequent complaint is stuffiness.

In order to provide enough ventilation when the VAV boxes are at their minimum settings, you really need to have airflow measuring stations on the outside air intakes. When meeting the outside air minimum airflow, if that causes the mixed air temperature to drop below the 55°F, it means you have to add a preheat coil. There goes some of savings these systems are supposed to provide.

So, to regain some comfort, many in-house personnel went to ridiculous efforts to reduce the minimum airflow to nearly zero. Insulation was stuffed into diffusers, VAV box minimums were turned to zero, etc. Not surprisingly, that led to even more indoor air quality issues. Away from the perimeter radiation, the only source of heat for occupied spaces was the internal heat gain from the people, lights and equipment. In most situations, it just wasn't sufficient.

I've never been a big fan of VAV systems. When cooling is desired, people are much more comfortable when they have larger quantities of moderate temperature air moving across their bodies. Increased volumes of air make it easier to properly ventilate too. Making that air supply a constant 55°F, year-round, doesn't come cheap either! Why make the supply air that cold, when something less cool will actually increase the comfort levels?

ASHRAE requires that all heating systems provide night setback capabilities. Sorry, turning the air-handling unit off, with the single temp thermostats all set at occupied temperatures, does nothing to reduce the unoccupied temperatures. Turning the supply fan off at night, only saves fan and ventilation energy.

Okay, so the industry responded with, "Let's figure out how to put some heat into both the interior and exterior of the building." In my opinion, what they came up with wasn't much of a solution. The new system simply blows the same constant 55°F air into a VAV box that sequences the modulation of a heating valve on a reheat coil, with the increased airflow for cooling.

I'm not certain anyone will ever be able to make a silk purse out of this sow's ear, but if anyone has a better solution, let me know.

Conclusion

Simultaneous cooling and heating is a little like setting the throttle on your car to the maximum RPM that you will ever need for accelerating from a dead stop, or to pass, then controlling the speed of your car with the brakes! While simultaneous cooling and heating makes about as much sense, I assure you, it happens all the time.

Whether the controls are pneumatic or electronic, it's pretty easy to determine if your HVAC system is providing simultaneous cooling and heating. If it's cold outside, look at the mixed air temperature. Again, that's the combination of outside air and return air. Is it 55°F to 60°F? It takes approximately 18% outside air, at -20°F, to maintain a 55°F mixed air temperature. But, that increases, linearly, to 100% outside air, at 55°F. If 18% can satisfy ventilation requirements, why bring in more, when it just adds to the amount of reheat needed and increases the potential for discomfort?

Remember, 55°F is the temperature of the air needed to cool the space at 90°F to 95°F outside air temperatures. Check the discharge air (DA) temp out of a few terminal boxes, compared to the DA temp of the air-handling unit. Is the system adding heat to keep that cold air from causing discomfort? Next, check back at various outside air temperatures. Is the unit DA temperature always 55°F? Are terminal boxes still adding heat? That's simultaneous cooling and heating. Another way to check is to turn the boiler(s) and heating water circulating pumps off. Does this cause discomfort at outside temps above 60°F? Again, if your mixed air, or air-conditioned air is always 55°F, you have simultaneous cooling and heating taking place. It's robbing you blind!

Air-handling systems are designed to cool, heat and ventilate. Today, however, we want the systems to only cool and ventilate, or heat and ventilate. To the fullest extent possible, you do not want all three at the same time. Now, there may be a few days a year, when the different building exposures, may require you to do all three at the same time. It's not the end of the world. Just be certain to know that, in fact, you need it and you keep it to a minimum!

Now, let's see how simultaneous cooling and heating impacts comfort. Everybody gives off a minimum 400 Btu's of heat an hour. If you have 30 students in a classroom, you have the equivalent of a 100% efficient, 12,000 Btu/hour furnace in the classroom. Then, add the heat from the lights, and perhaps even a little solar gain, and the point at which the classroom internal heat is finally less than the external heat loss, is about +10°F outside air temperature. That means, after initial morning warm-up, the typical classroom is overheated any time the outside air temperature is +10°F, or warmer. I've also had an opportunity to work with a small casino. Even with large minimum amounts of outside air for ventilation of smoking areas, the point at which their air-handling units start to use free atmospheric cooling is about 35°F. The heat is off at even lower outside air temperatures.

This situation then becomes a real comfort problem, when someone wants to save heating energy, like when Jimmy Carter said that we couldn't use additional heating energy to heat space temperatures above 65°F? Nor could we use cooling energy to cool spaces below 78°F? Back then, most HVAC systems were some form of reheat system. Instead of helping to solve the energy crisis, with temperature control system revisions, all that was done in most facilities during the winter, was to turn the thermostats down to 65°F. With simultaneous cooling and heating, when the thermostat turns off the heat, because of internal heat gain, everyone blamed Jimmy for all the cold air blowing at them!

In the summer months, again, all that was done, in thousands of facilities, all across America, was to turn the thermostats up to 78°F. With the heating sources still fully operational, space temperatures were heated to 78°F. How comfortable was that? Then, all that additional heat was cooled back down to 55°F, before being reheated, again. Both heating and cooling costs increased and Jimmy couldn't get re-elected, because he made all of us unnecessarily uncomfortable! Jimmy – you should have asked me for some help.

All this adds to why it's good to have the big picture in mind, when you start a maintenance recovery program. It's not that you can't eliminate simultaneous cooling and heating, with an antiquated pneumatic temperature control system. But, finding someone that understands how to do it, then be there to maintain it well into the future, probably isn't in the cards. Now, that's not to say that because someone is installing a state-of-the-art building automation system, BAS, it means that the new system won't be providing the same simultaneous cooling and heating. If the contractor doesn't understand how to do it with pneumatic controls, they are not likely to understand how to implement efficient sequences with a BAS. I highly recommend that you interview perspective contractors and ask a lot of questions. Use the above discussions to help you design your questions.

There are many different styles of air-handling systems. These are just a few samples of how a system can provide energy wasting, comfort robbing, simultaneous cooling and heating. There are other HVAC systems that are so new, I don't have any experience. They include displacement ventilation, chilled beam systems and more. If considering use of any new system design, I'd recommend finding an engineer that has designed more than one. Check with the owners to see what works and what doesn't. Remember to look out for cold air that gets reheated. It's not rocket science!

Temperature control system revisions

4.B – Enthalpy control

Enthalpy is the total heat content of the air, measuring both temperature and humidity. During the air-conditioning season, it's most efficient to cool the air with the lowest total heat content. So, most HVAC systems go to great lengths to compare the enthalpy of the outside air to the enthalpy of the return air, again, selecting the one with the lowest enthalpy.

As you've seen in the Chapter 4.A discussions, many of the early HVAC designs used a constant 55°F discharge air-conditioning temperature. So, when the outside air temperature exceeded 55°F, the outside air would be reduced to the minimum quantity required for ventilation. This temperature is called the dry bulb switchover temperature.

Then, when computers became commonplace, someone with too much time on their hands, entered 10 years' worth of climatological data into one and compared it to the typical return air condition of 75°F/50% relative humidity for a space temp of 72°F. Low and behold, in many areas of the country, you could save lots of money by allowing the outside air to be used higher outside air temps. This started the frenzy to create, and sell, enthalpy control.

At that time, I was working for a major, national temperature control company. In their efforts to help the sales staff sell enthalpy control, they created a chart that showed the savings of a dry bulb switchover, compared to "true" enthalpy control. The chart showed the savings potential by using "true" enthalpy control, versus a dry bulb switchover, at different temps. The chart for Minneapolis, MN was as follows:

Outside switchover temperature	Savings with "true" enthalpy
55°F	21.96%
60°F	12.77%
70°F	3.28%
75°F	7.89%

The state-of-the-art in enthalpy control, at the time ~ 1975, was demonstrating that it was less accurate, and less repeatable, than the savings from a simple 70°F dry bulb switchover, so I called and talked to the engineer that did the analysis. I asked if he just ran the comparisons at 5°F increments, or if they looked at 1°F increments. When he said 1°F, I asked, at what temp did the dry bulb switchover equal 0% savings, compared to "true" enthalpy control? The answer for Minneapolis was 69°F. With that info, I stopped promoting enthalpy control, **until now**.

I recently had an experience where the enthalpy controls on each of 6 large rooftop units, all serving a huge common space, were not switching over at anywhere near the same outside conditions. I will refrain from mentioning any brand name, but they retail at Grainger for $111.95. They use some measurement of outside air enthalpy, compared to an assumed return air condition of 75°F/50% relative humidity. So, I tried using a dry bulb switchover, set at 69°F.

All worked well, until the two days I wasn't there and it was raining. The outside air temp was consistently between 63°F and 68°F. The occupants of this 24 hour-a-day operation were not very happy!

Bottom line, I'm convinced of the need for the best enthalpy control available. However, I'm not convinced that each and every air-handling unit needs their own enthalpy control. I believe you can pick a representative, or most critical, return air enthalpy and compare it to the outside air enthalpy. Then, use that one, very accurate switchover condition, for all building air-handling units.

Now, accurate, repeatable enthalpy control is expensive and I'm always trying to keep costs low, so if your indoor environment has more critical operating conditions, or if your budget allows, please feel free to provide a high quality, "true" enthalpy control for each air-handling unit.

One note of caution – be certain to ask your control contractor if the humidity sensor needs to be taken inside, when the outside air temperatures fall below 32°F. Some computer chip humidity sensors can be ruined if they freeze.

Temperature control system revisions

4.C – Natatorium air-handling unit (Swimming pool enclosure unit)

In my discussion of air-handling systems for space cooling and heating, I said that their purpose was to heat, ventilate and cool. When it comes to natatorium air-handling systems, we also have to add dehumidification to our list of tasks to be performed. For purposes of this discussion, I'm going old school for natatorium dehumidification. That is, I'm going to discuss what I call traditional dehumidification air-handling unit design. Because I intend to demonstrate superior cost-effectiveness, I will only mention mechanical dehumidification systems in passing.

I want to start out by clearing up a potential area of confusion. In the ASHRAE Handbook on HVAC Systems and Applications, they state that natatoriums should be maintained at 55% relative humidity (RH). At least one engineer, in the area where my business operated, really didn't understand that natatoriums only have a problem with too much humidity, not too little. On three separate natatoriums, in the same school district, I personally turned off the steam supply to natatorium space humidifiers!

When considering a dehumidification system for a natatorium, the first consideration is whether or not the space ever needs to be cooled. Pool water temperatures are always supposed to be between 80°F and 82°F. In order to keep the swimmers comfortable while nearly naked, the space temperature needs to be 2°F warmer than the water. Now, let's say that it's 95°F outside. What do most people want to do on a day like that? Right, they put on a bathing suit and get wet. With evaporative cooling taking place, as soon as one gets out of the water, is there really a need to blow cold air on that person? Would it be necessary to blow cold air on someone, if they were swimming outside on a 95°F day? For me, the answer is no, but that's the basic question you need to answer for your particular situation. A properly designed air-handling system should be able to keep space temperatures within a couple degrees of the outside air temperature. Possible exceptions to no need for cooling would be:

1. Health clubs, where the mostly adult clients demand greater comfort.
2. Natatoriums having large spectator audiences, also requiring greater comfort.

But, since these are limited applications, and you could just add some air-conditioning for those special times, I will proceed with the discussion of traditional air-handling systems. I believe that, if there is no need for artificial cooling, there is no justification for mechanical dehumidification. Mechanical dehumidification exists because the design engineers don't fully understand proper temperature and humidity control strategies of traditional natatorium air-handling systems.

The premise of being able to dehumidify, without mechanical dehumidification, is based on the fact that, when air is heated 10°F, the relative humidity is reduced by approximately 30%. So, at least, in the northern areas of the United States, where the outside air is cooler than the inside air most of the year, it makes sense that we bring in more outside air, and simply heat it up, when we need to dehumidify.

The ASHRAE Handbook on HVAC Systems and Applications, describes the air-handling unit for natatoriums as requiring the capability of providing 100% outside air for dehumidification purposes. It also suggests that, beyond a minimum amount of outside air, which may vary from state to state, a humidity controller be used to control the introduction of outside air. In climates where the outside air temperature never drops below 45°F, I have no problems with this limited description of the sequence of operation. But, in all other colder locations, that sequence of operation requires considerable additional discussion.

One of the biggest problems associated with natatorium air-handling systems, begins at outside air temperatures below 32°F. With the warm, moist return air, combining with the cold outside air, there will likely be a frost build-up on the air filters. Occasionally, this frost build-up is so severe, the pressure drop across the filters exceeds the structural capabilities of the filters and they collapse. With some proper planning, regarding the space required for proper mixing ahead of the filters, a stationary air-blending unit would thoroughly mix the outside air and return air to help alleviate this problem.

The stationary air-blending unit will only help solve this problem, however, if the minimum, or humidity controlled quantity of outside air, does not drive the resulting mixed air temperature below the freezing point. So, we need to determine if the quantity of outside air, required to meet both outside air minimums and dehumidification requirements, could be sufficiently high enough to keep the mixed air temperature above 32°F.

The minimum ventilation requirement for swimming pools, in Minnesota, is 0.5 cubic feet, CF, of outside air for every square foot, SF, of pool and deck area. This code used to say, pool and wetted deck area, and I suppose that could still be argued, but using the entire pool and deck area presents a worst case condition, so I'll us that.

Let's start with a typical 50' X 75' pool, with a 20' deck all the way around it. That's 90' X 115' or 10,350 square feet, which requires 5,175 cubic feet per minute, CFM, of outside air, as a minimum. I have seen nothing in the

Minnesota, or the International Building Code, that suggests that mechanical dehumidification systems can provide anything less. The minimum is the minimum! That's going to be important before we're done.

Most natatoriums have about 20' ceilings. That makes our example natatorium 207,000 CF. ASHRAE suggest that natatorium air-handling units be designed with an air turnover rate of six times per hour. That means that the entire volume of 207,000 cubic feet of air be moved through the air-handling unit, six times every hour, or in this example, it be sized at 20,700 CFM.

Now, it's a simple calculation to determine the mixed air temperature, at design outside conditions. 5,175 CFM of minimum outside air, divided by 20,700 CFM of total air supply, yields a minimum percent of outside air at 25%. Using -20°F design outside air temperature for a Minnesota winter, and 78°F, as the absolute coldest space temperature, the following formula will yield the worst case minimum mixed air temperature, when satisfying minimum ventilation requirements:

(-20 OAT x 25% OA) + (78 RAT x 75% RA) = 100% MA x MAT = 53.5°F = Minimum mixed air temperature, when satisfying Minnesota's minimum ventilation requirements. Even if it ever gets down to -30°F, the minimum mixed air temperature is still 51°F. Again, check with your state to see if outside minimums for natatoriums differ.

OAT = Outside Air Temperature
%OA = Percent Outside Air
RAT = Return Air Temperature
%RA = Percent Return Air
%MA = Percent Mixed Air
MAT = Mixed Air Temperature

Well, 53.5°F is certainly well above the freezing point. So, in terms of fulfilling outside air minimums, there is certainly no reason for allowing the mixed air temperature to drop below the problem temperature of 32°F. Now, we need to determine if a 53.5°F mixed air temperature will provide the necessary humidity control.

The design humidity condition for a natatorium is 55% RH, plus or minus 5%. Again for a worst case scenario, let's say that we're bringing in 100% outside air, at 50°F (I'm rounding down to 50°F to make life easier for me.) and the RH is 100%. Remembering that a 10°F rise is temperature reduces the RH by approximately 30%, heating the supply air to 60°F, the RH is reduced to approximately 70%. Heat it to 70°F and the RH is lowered to approximately 49%. Heat it 10°F, once more, to the minimum discharge temperature allowed by a properly operating natatorium air-handling unit, 80°F, and the

RH of 20,700 CFM of supply air drops to approximately 34% RH. Experience has repeatedly illustrated that this quantity of 34% RH air, will dry any natatorium to 55% RH. If you want to be even more conservative, you could easily lower the minimum mixed air temperature to 45°F or even 40°F. (For those with more experience using a psychometric chart, I'm certain you'll confirm what I've just presented.)

So, I've gone to great length to prove that a 50°F mixed air temperature met both the minimum ventilation rate requirements and dehumidification requirements, at the coldest of outside temperatures. Why is that so important? Well, remember back in the beginning when it was noted that ASHRAE said that a natatorium air-handling unit had to have the capability of providing 100% outside air? I agree that these units be designed to go to 100%, but when an air-handling unit is driven to 100% outside air, at extreme outside air design temperatures, the previously mentioned frosted air filter problem arises. Such an operation also causes much more hardware to be installed, such as multiple preheat coils and huge amounts of additional heating energy, to heat the space and keep coils from freezing. Instead, the proper installation of an air blender would eliminate all that cost and energy waste. The contact info for the specific air blender that I've used will be shown at the end of this chapter.

The ASHRAE Handbook also discusses the need to control exhaust from the natatorium, so that a negative pressure is maintained. This is needed to alleviate the potential problem of moisture in the building envelope. A negative pressure also helps keep the chlorine odor from circulating to other parts of the building, which causes accelerated deterioration of many metal items.

If operated at a positive pressure, no matter what kind of vapor barrier you might have, the high humidity air will penetrate the outside wall and roof structure. And, just like the humidity going down when air is heated, humidity goes up when air is cooled. During the heating season, the air passing into the wall or roof will be cooled to a dewpoint temperature of approximately 63°F. Then, when it gets really cold, that moisture can freeze, causing considerable damage to the bricks, mortar and roof structure, as it expands. Exhaust capabilities must exceed outside air intake capabilities. Variable frequency drives (VFD's), controlled by building static pressure, are the solution.

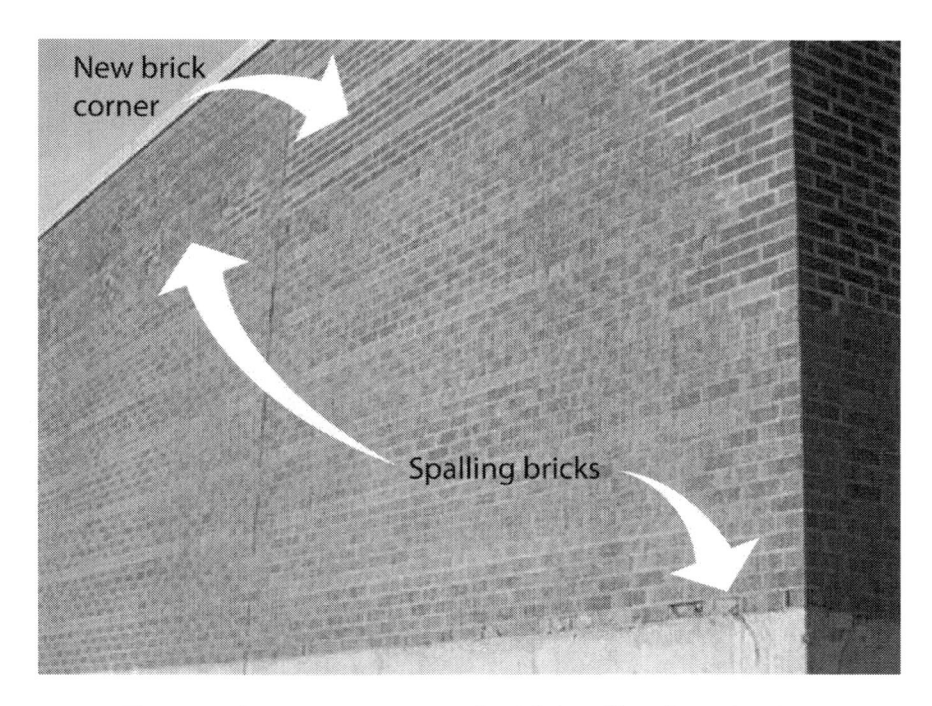

Figure 4.C.1 Natatorium wall with spalling from having a positive pressure.

In the above photo, from the roof down several feet, and in the lower right-hand corner, you can see that many bricks have lost the front portion of the bricks. That's called spalling. It's caused by the pool moisture penetrating the wall. The exhaust is not exceeding the incoming outside air. Each spring, they had to walk the perimeter to pick up brick halves before they could mow the lawn! In the upper right-hand corner, you can see where a large portion of the building envelope had previously been replaced, because it looked like the corner in the following picture.

Figure 4.C.2 Natatorium walls falling apart because of a positive pressure.

Eventually, the entire exterior envelope of this natatorium had to be replaced, costing hundreds of thousands of dollars and they still had to solve the problem. It is not an uncommon occurrence. I've even heard of one natatorium roof that collapsed in one southern Minnesota school. Fortunately, the natatorium was unoccupied. If you can smell the pool chlorine anywhere in the building connected to the pool, the natatorium has a positive pressure and you too will eventually be dealing with a similar situation. Fixing it early is the key. If necessary, share this chapter with your control contractor.

Now, let's talk about pool blankets. First of all, automatic pool blankets that go on and off with the turn of a key, should eliminate most of the arguments about who is going to put the blanket on and take it off. ASHRAE says that, if you have mechanical dehumidification, you don't need a pool blanket. My question is, if you are regularly using a pool blanket, and are providing the required minimum amount of outside air when the blanket is off, is there enough economic justification for the high cost of installing and operating a mechanical dehumidification systems? If the analysis looks like it may be a close call, don't forget to throw in the high cost of maintaining, and/or replacing, the air-conditioning compressors used in the mechanical dehumidification process. See Chapter 5.G for a photo of an automatic pool blanket, along with contact info for the manufacturer.

The final item for natatorium air-handling unit control is the night cycle. Unless the pool is being covered with a blanket, cycling the air-handling system at a reduced night setback temperature should never occur. As the space temperature drops, the percent relative humidity increases significantly. With the considerable heat sink of the pool water, it will likely rain in the natatorium, before a reduced night setback temperature is ever achieved. Therefore, the night cycle operation, with the use of a pool blanket, should be controlled by either a modest couple degree drop in temperature, or an increase in the relative humidity to about 60% RH, whichever occurs first. The humidity control will also assure proper operation, should someone forget to put the blanket on. Don't forget, maintaining a negative pressure has to be 24/7!

In summary, both outside air minimums, and humidity control, can be provided in cold weather climates, without the problems, and expense, associated with allowing the mixed air temperatures to be less than 32°F. Blending outside air and return air, keeping the mixed air temperature at or above 50°F:

1. Eliminates the frost build-up problem on the filters
2. Eliminates the size and quantity of the preheat coils and control valves
3. Eliminates the need for face and by-pass damper control, to keep from overheating caused by control valves that are wide open, trying to keep their coils from freezing
4. Eliminates the need for huge amounts of steam or glycol treated hot water to keep heating coils from freezing

Again, if your natatorium doesn't need mechanical cooling to keep spectators comfortable, or if you're just not willing to pay the expense to make that happen, be certain to have your engineer consider the information of this chapter when making the economic comparison of traditional, versus mechanical dehumidification systems.

One final item on this subject. Retrofitting a natatorium air-handling system can give you quite an opportunity to use your mechanical skills. On one such occasion, a school had a 100% OA air-handling unit with: two preheat coils; face and bypass dampers; and one reheat coil. The outside air came in over the top of the air-handling unit and made a U-turn, back in the direction from which it came, to go through the coils.

About 15' away, they had a 100% return air, RA, air-handling unit. Like the other unit, the air came in above the unit, made a U-turn into the unit, which blew the air back in the direction from which it came. This unit had no heating capabilities. As they sat, the units' output blew in opposite directions. Both air-handling units were on the same plane. I know, it sounds like a pretty screwy design, but I had a plan.

After disconnecting the ductwork from the 100% RA unit, it was lifted and turned 180°, so that the RA could be blown into the other unit. A damper was installed on this unit.

The OA ductwork from the 100% OA unit was extended to the RA unit, where an air blender was installed as far from the filter bank as we could get it. If I recall, the air blender needs 4 times the smallest duct diameter for proper blending. We had more than enough space. On this unit, all preheat coils were removed. This work was before we knew about automatic pool blankets. A small BAS was installed to control both air-handling units and the exhaust fans, as described above.

I hope this helps you, as you look to redesign your old natatorium air-handling system or consider a new natatorium installation.

The air blender manufacturer with which I am familiar, mentioned above, is:

Blender Products, Inc.
5010 Cook Street
Denver, CO 80216

800-523-5707
http://www.airblender.com

Install new technology hardware

5.A – Best Value procurement

I started out to describe a method of procurement my company referred to as Quality Based Selection, QBS. QBS is a different method of procurement than the plan, specify and bid procedure, with which you are probably most familiar. As I struggled with the details, I checked with the current owner to see if she had something that could help. Always looking for third party validation, what she sent back was a link to the Arizona State University (ASU), Performance Based Studies Research Group. Instead of just looking to have the least costly project, QBS, or as ASU calls it, Best Value, helps you get the best valued project. Actually, what it does is help you select the most qualified contractor, instead of just the cheapest. As a contractor that prided ourselves on the best performing work, we promoted QBS at every turn. In Minnesota, QBS/Best Value is becoming a more commonly selected procurement method. Of course, private building owners have no limitations on their methods of procurement. With improved quality, I believe you will soon come to rely on Best Value for all future procurements.

So, I just erased my initial write-up about QBS, and am, instead, directing you to the ASU website on Best Value at: www.pbsrg.com.

As you view the videos at this website, their examples are about some very large projects. You should know that the process is applicable to smaller sized projects, as well.

I've been looking for a place to add this little gem to the discussion and this seems appropriate. "Price, Quality, Service. Choose Two." I know you've all heard it. What Best Value is telling you is that quality and service are the most important and you get to negotiate a fair and reasonable price. This allows both the owner and the contractor to have good expectations for the project. My bet is that, at the end of this type of project, if the contractor wasn't one that you knew, liked and trusted before you started, they will be.

Check out the above website before making any major purchases. I'm confident that you'll like the outcome.

Install new technology hardware

5.B – Boilers

Steam

Steam heating should be allowed to die. Should you even want to make an addition to a steam heated building, and the engineer wants to simply extend the steam system, fire that engineer, immediately! I once found a school that extended the steam to a significant addition. I knew the engineer, so I called and asked why. He said that's what the district wanted. Sorry, you don't hire engineers to blindly do what the uninformed school board tells the engineer to do. At least, not without considerable sharing of information. A good engineer will help keep the owner from making costly mistakes, like extending an existing steam system. The boilers were already past their published life expectancy. Extending the steam system should have never happened.

A couple years later, shortly after my retirement, my firm finally convinced the superintendent and school board of the benefits of converting to hot water. They were also convinced of the benefit of the QBS process, described in Chapter 5.A. I had, apparently, demonstrated some level of competence, during the time I was working with the district, so they asked me if I would help them with the contractor selection process, if needed. Like is mentioned in the ASU Best Value discussion, the most qualified contractor usually presents itself. I was not present at the contractor presentations, but the superintendent called the next morning to inform me that my services would not be needed. My former company's proposal was clearly superior and selected.

I recently checked with the project manager and learned that the air-handling systems were also updated to eliminate the simultaneous cooling and heating, so an exact savings from the conversion to hot water is not possible. However, the energy savings from the entire project was approximately 45%!

Okay, let's say that your steam boiler is absolutely in need of replacement. You've made it stretch nearly twice its 25 to 35 year life expectancy. The first response of many people would be to simply order a new steam boiler. The problem is that this pretty much sets the heating plant in an inappropriate direction for another 25 to 50 years. If you later find that the steam pipes need replacement, most steam boilers can be converted to hot water, but you'll never attain the efficiencies of a modern condensing boiler.

Instead of just installing a new steam boiler, I encourage you to seriously look at the age and condition of your steam boilers **and pipes**. If replacement of either, or both, is imminent, plan ahead for it. A project of this magnitude must be planned up to a year in advance. If you're a commercial facility, you've just got the non-heating months. If you're a school, you've got even less time. When looking at replacing steam heating coils, with hot water heating coils, it is also an excellent time to upgrade the air-handling system to provide the proper ventilation and eliminate simultaneous cooling and heating. Whether you just replace the coils; look at completely new air-handling systems; or also add a new building automation system, BAS, all require long-term planning.

Now let's talk about what to do when you can't convert to hot water heat, but you have to improve the efficiency of the boiler. Many of the old boilers, with the large burners hanging on the front, can improve efficiency by installing the most efficient burner available.

Figure 5.B.1 Example of Weishaupt burner, with high turndown ratio.

With the reduction in energy consumption that you've already received from the first four steps of this book, you may well be able to consider installing a smaller burner. In addition, the lowest firing rate of your old burner may be as high as 50% of its maximum firing rate. Another way of looking at this is that the old burner only has a 2:1 turndown ratio. This means that at low heating needs, your boiler is cycling on and off a lot more than necessary.

Firing rates for new burners can be as low as 20% of its maximum capacity, or a 5:1 turndown ratio. Combined with a smaller size burner that still works with the size boiler you have, it can mean much longer, more efficient run times.

Turndown and Why it's Important

(Thanks to Brett Stueland of the RM Cotton Co., the MN manufacturer's rep for Aerco Boilers and Riello Burners, for the following discussion.)

"The turndown ratio has a significant effect on system performance; lack of consideration of the source system's part-load capability has been responsible for many systems that either do not function properly or do so at the expense of excess energy consumption." – 2012 ASHRAE Handbook

Turndown is an indication of the minimum firing rate of a burner/boiler. It is simply a measurement of how low the firing rate can be reduced ("turned down") from high fire. Turndown is expressed as a ratio = Maximum Firing Rate / Minimum Firing Rate.

For example, a boiler with a maximum firing rate of 2000MBH and a minimum firing rate of 100MBH would have a turndown ratio of 20:1 (2000/100 = 20). A boiler with a maximum firing rate of 2000MBH but only a minimum firing rate of 400MBH would only have a 5:1 turndown. As you can see, a higher turndown will allow the burner/boiler to operate at a lower firing rate. Why is this important?

Having the capability to operate at lower firing rates allows the burner to stay on and modulate to match load changes in lieu of cycling off and on. This cycling is even more profound at part load conditions because boilers are almost always oversized. On the following chart, the line made up of small pieces of straight lines is the building load line. As you can see from this chart, under part load conditions a higher turndown burner, 20:1, will more closely match the load, whereas a lower turndown burner 5:1, will cycle off and on. High turndown is beneficial for a variety of reasons. In order to understand some of these benefits, it is important to know how a burner operates.

BTU — 5:1 Burner, 20:1 Burner

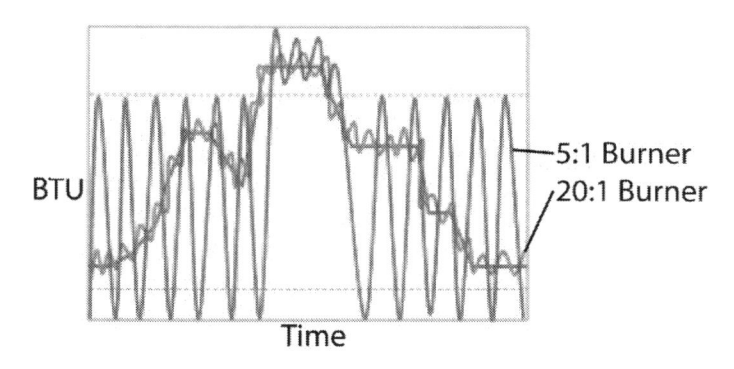

Time

In simple terms, a temperature or pressure sensor enables a burner to fire. When the burner receives this signal to fire, it must first go through a series of safety checks. These checks could include proving the following: water flow, combustion air damper open, gas valves open, and various pressure switches (blocked filters, high gas pressure, low gas pressure, combustion air flow). A burner may have some or all additional checks depending on the application and size of the burner. Furthermore, all burners have a pre and post purge safety feature. Pre purge energizes the fan on the burner prior to allowing it to fire in order to purge any possible residual gas left in the firing chamber. Post purge energizes the fan after the burner has fired. Each of these purges can run from 15 to 90 seconds. Once the burner has gone through all of its safety checks and pre purge, it is then allowed to fire. A modulating burner will adjust the firing rate as needed to maintain a given temperature or pressure setpoint. As the water temperature or steam pressure rises to near the setpoint, the burner will reduce its firing rate.

When the building load is reduced to a point that is below the minimum firing rate of the burner, the water temperature or steam pressure will rise above the setpoint and the burner will shut off and start its post purge. In a worst case scenario, the water temperature or steam pressure will drop quickly and the burner will start again, and again go through all its pre-fire checks.

This on and off cycling leads to many inefficiencies. The time it takes to execute the purges and other safety checks can delay the burner from firing which can lead to unwanted fluctuations in temperature or pressure. Each time a burner purges it steals heat/energy from the boiler and discards it out the stack. Even when the burner is in its off cycle, additional energy is lost due to the natural drafting stack effect. Furthermore, cycling causes greater wear and tear on electrical and mechanical components as well as increased stress due to thermal expansion and contraction.

"A higher turndown ratio reduces burner starts, provides better load control, saves wear and tear on the burner, reduces refractory wear, reduces purge-air requirements, and provides fuel savings" – US Department of Energy, Upgrade Boilers with Energy-Efficient Burners article.

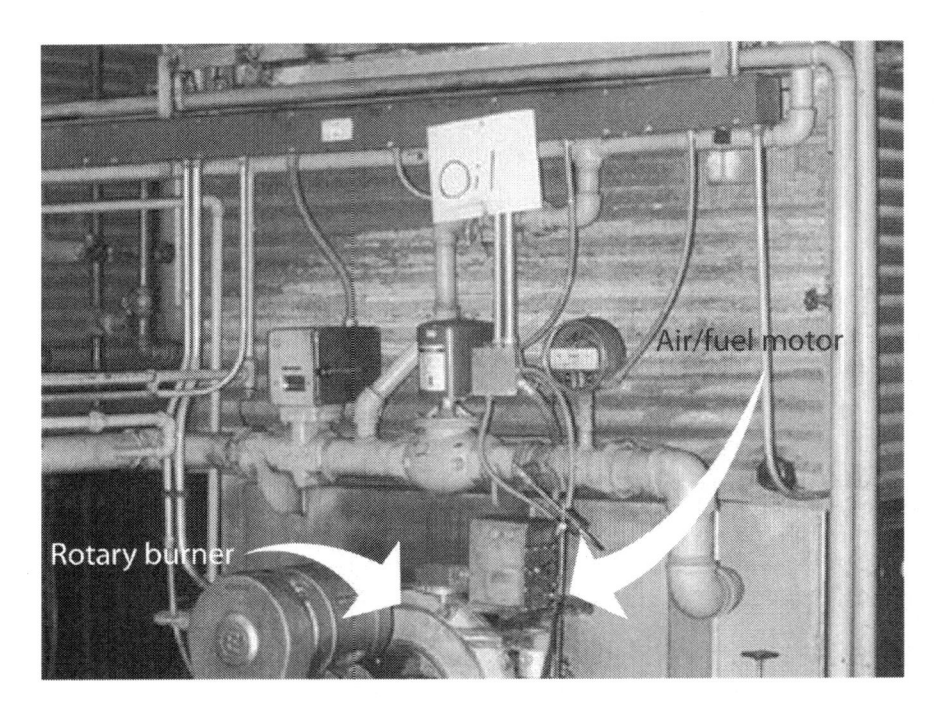

Figure 5.B.2 Big box steam boiler, with an old rotary burner.

This boiler, and the identical backup boiler sitting next to it, are classic examples of boilers still in operation, well after their 25 to 35 year life expectancy. See the gray box, above the burner, below the pipe? That's the motor that mixes air with the fuel. Combustion efficiency depends on the proper adjustment of the rods connected to the gas valve and combustion damper. Getting maximum efficiency over the entire range is difficult, at best. One bump that moves any connection can cost you lots of money.

I certainly wouldn't recommend this for the above pictured burner, but another new technology improvement to burners is linkageless burner control. It can be added to old burners to improve their efficiency anywhere from 3% to 10%. It can also be factory installed on new burners. Use the energy saving estimate from your burner service company in the benchmark analysis in Chapter 6.D, to help determine the return on investment, ROI.

Hot water

On a seasonal basis, most of today's steam boilers are less than 70% efficient. The long-term benefits of operating with 92% seasonally efficient, condensing boilers should be obvious. I'll discuss how to achieve that level of efficiency more in Chapter 5.C. However, the payback is highly dependent on the cost of fuel. At one time, dual fuel natural gas rates in Minnesota were

$1.00/Therm. In August, 2014, the rate was less than $0.50/Therm. That makes payback more difficult, but I'm confident the advice being provided will help maximize the ROI. No matter what the gas rate, the benchmark analysis of Chapter 6.D can help you determine the ROI.

Condensing hot water boilers are the only style that I recommend. At least one manufacturer touts a 20:1 turndown ratio, which means it can provide a steady, efficient runtime, anytime the heating load is 5% of the maximum capacity, or more. With multiple boilers, that steady runtime can be lowered even further. But, if such boilers are not within your budget capabilities, then a new burner, with the addition of linkageless burner control, may be an excellent option. Again, the benchmark analysis of Chapter 6.D can help you determine the proper size.

What you're constantly trying to do is have a condensing boiler operate with return water temperatures at, or below the condensing temperature of 130°F. See how significantly the efficiency increases, as the return water temperature drops below that value.

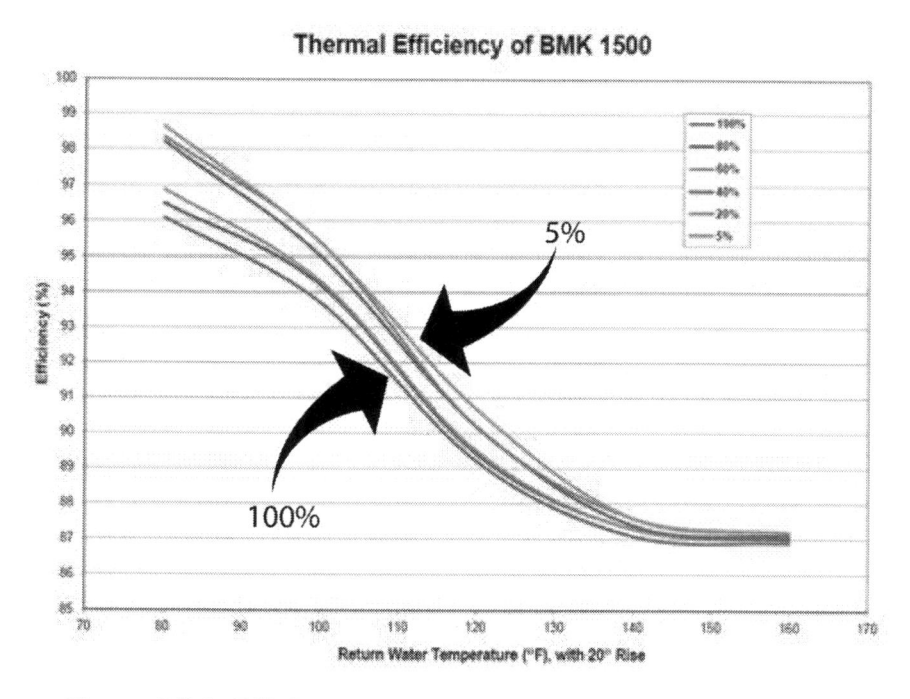

Figure 5.B.3 Efficiency curve for an Aerco condensing boiler.

Also note how the efficiency goes up with the boiler operating at part loads. This is why it's important to have multiple boilers that can operate intelligently. At the end of this chapter, you'll find links to Aerco documents that describe how they operate multiple boilers in part load conditions. They will also describe the benefits of using isolation valves to assure that short cycling doesn't occur from low flow. You will get much more precise control when you're not trying to satisfy a single sensor at the end of a long row of boilers, all having partial flow.

On that last subject, a school for which my former company provides repair and maintenance services, has 10 low efficiency, noncondensing boilers in a row. The outdoor reset control tries to sequence the boilers to maintain the setpoint temperature, at a single sensor, located downstream of the last boiler. The main problem that they have is that the first boiler(s) turn off on high limit, long before satisfying the setpoint. You have to like a company that knows enough to solve that problem before it occurs.

Figure 5.B.4 Modern, efficient condensing boilers.

The above boilers are examples of small condensing boilers that can have a 20:1 turndown ratio. They can accept an external signal from a BAS to operate at any water temperature, as described in the improved hot water control discussion of Chapter 5.C.

When planning your heating system conversion, from steam to hot water, be certain to include analyzing the benefit of having an interruptible fuel supply, as well as the improved hot water reset control that follows shortly. The February, 2014 price differential, between firm and interruptible gas in Minnesota was about 15.4%. I consulted with one of the boiler specialists at my old company about dual fuel condensing boilers. His thoughts were that if the building had an existing backup propane system, the added cost of dual fuel would be worth it. If not, a lot would depend on the size of your facility; the efficiency of your new boilers; and the price difference between firm and interruptible natural gas, at the time.

Now, a couple words about backup fuel supplies. Remember the days when underground tanks, with 10,000 gallons of fuel oil, were the normal backup fuel supply? When checking the above interruptible pricing difference, I also found a recommendation that the backup fuel supply be just 20% of the average January consumption. That's pretty reasonable!

In the winter of 2013/2014, Minnesota was caught in the polar vortex, causing them to experience 53 days with below zero temperatures. It had been 7 years since interruptible natural gas customers had experienced any interruptions. Needless to say, the boiler service companies were extremely busy getting backup systems to work. Make certain your boilers are ready to operate on the backup fuel supply, no matter what it is.

Primary/secondary vs. variable primary piping for boilers

(Thanks, again, to Brett Stueland of the RM Cotton Co., the MN manufacturer's rep for Aerco Boilers and Riello Burners, for the following discussion.)

Primary/Secondary piping was originally developed for use with copper finned watertube boilers when they were first introduced years ago. As these "new" watertube boilers were a low mass/low water volume design, they needed a method to ensure proper flow to prevent overheating and scale buildup. And they needed to handle the much higher pressure drop associated with watertube boilers. Primary/Secondary piping solved all of those issues.

Primary/secondary piping consists of two loops each having its own pump, where the main loop is typically referred to as the secondary and the boiler loop is referred to as the primary. With each boiler having its own pump, the proper flow is achieved. Furthermore, the primary loop is decoupled from the secondary loop thereby removing the high boiler pressure drop from the secondary loop. Figure 1 shows basic layouts for a single boiler and multiple boilers.

Figure 1

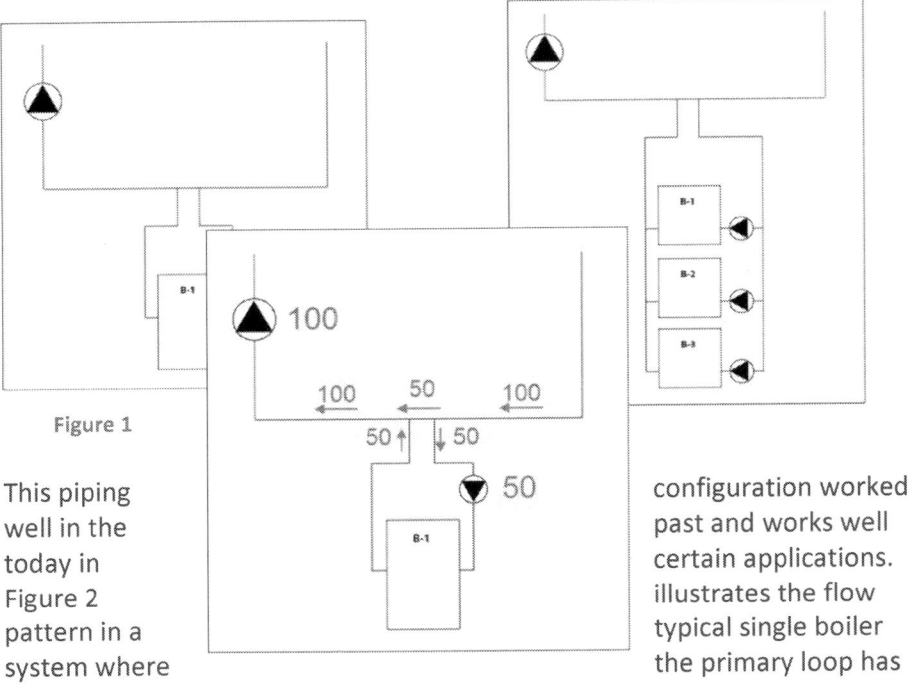

This piping configuration worked well in the past and works well today in certain applications. Figure 2 illustrates the flow pattern in a typical single boiler system where the primary loop has less flow than the secondary. Note that all the heated water leaving the boiler enters the main loop and out into the system. This is great flow for the boiler. The only drawback would be that the heated water entering the main loop actually gets mixed with cooler water that bypasses the boiler. This may cause the boiler to run at higher temperatures but eventually the loop temperature would equalize.

Figure 2

Today, energy conservation is very important and most applications will have a Variable Frequency Drive (VFD) on the main loop pump. Figure 3 now shows the flow pattern when the pump is on low speed assuming about a 1/3 flow reduction when using a VFD. Note that all the heated water no longer goes out to the system. The flow actually reverses and recirculates warm water back to the boiler. This is not desirable for a two reasons. First, warmer returning to the boiler reduces its efficiency. See Figure 4 – Boiler Efficiency Chart based on Return Water Temp from ASHRAE handbook. Second, warmer return water increases boiler cycling and can possibly trip the boiler high temperature limit. Excessive cycling creates more fatigue on the boiler and could void the boiler warranty.

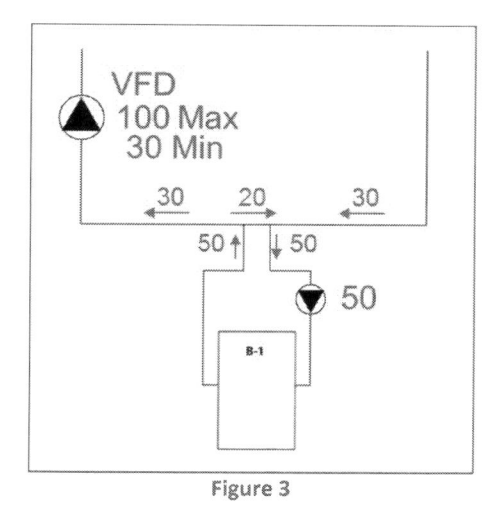

Figure 3

Figure 4

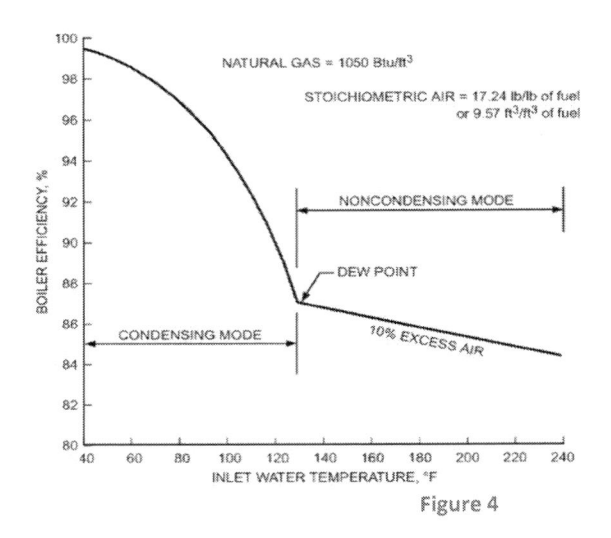

There are options to consider to help reduce these issues like limiting the VFD, using buffer tanks, or adding more boilers to reduce the individual boiler flow. But these defeat the purpose, slow down the inevitable and/or add cost the project. Perhaps a better option would be to use a Full Flow or Variable Primary design.

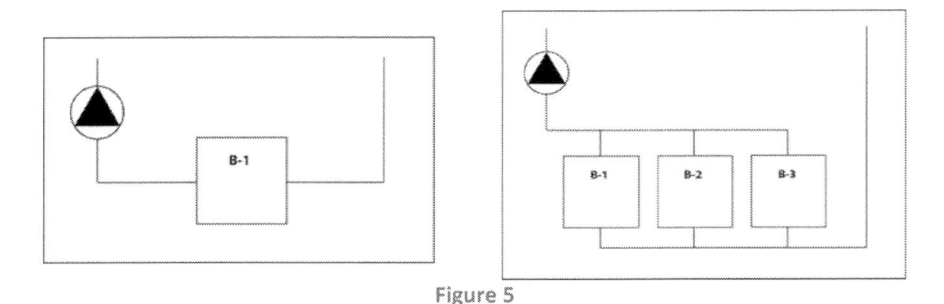

Figure 5

A Full Flow system uses the main loop pump for flow through the boiler(s). And unlike primary/secondary, no boiler pumps are used. Figure 5 shows a typical drawing for a single boiler and one for multiple boilers. Prior to the introduction of low mass boilers, this was how all boilers were piped. The only concerns with this piping system are to maintain the minimum flow rate of the boiler(s) and to balance the flow between the boilers. Motorized isolation valves can be used to ensure each boiler receives its required minimum flow. Figure 6 shows a drawing using motorized isolation valves.

Figure 6

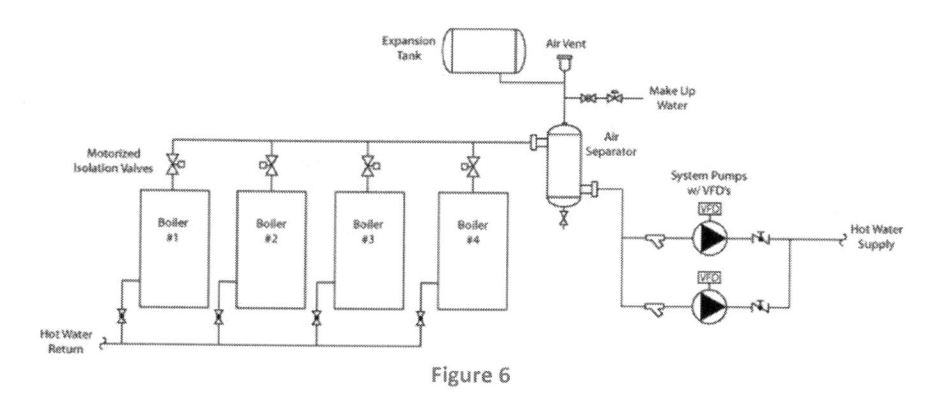

Each motorized isolation valve is only open when its respective boiler is on. Therefore, if only one boiler is on, the system pumps will be at low flow and all the water will be forced through the on boiler. If there were no valves the low flow would be divided through all the boilers which could result in flow rates being below the minimum requirement of the boiler. This would cause excessive cycling. In addition, if all the boilers are off, one or more valves must be open to prevent dead heading the pumps. The control of these valves can be performed by the Building Management System or maybe more preferably, by the boiler controller. All Aerco Benchmark boilers now have an integral lead/lag sequencer with motorized isolation valve control.

In conclusion, the full flow or variable primary piping does not have the issues associated with primary/secondary. All boilers see the coldest possible temperature for maximum efficiency and reduced boiler cycling. And as there are no boiler pumps and the valves are only open/close, less electricity is consumed.

Electric boilers

In some areas of the mid-west, I have found rural electric associations, REA's, have some very attractive electric rates, without demand charges. If your REA has such rates, it might be interesting to see how a nearly 100% efficient electric boiler compares to the cost of operation of a 92% efficient boiler. Plug the numbers into the benchmark analysis in Chapter 6.D and see. We did this for a modest sized K-12 school, back when natural gas was about $1.00/Therm. The client had a long term commitment from the REA, but we didn't know that natural gas was going to go to $0.50/Therm. As smart as it looked at the time, it's still a bit of a crap shoot! Just look at all options.

Another well recommended burner is Riello, made in Italy.

Riello Burners North America
35 Pond Park RD
Hingham, MA 02043
781-749-8292
www.riello-burners.com
info@riello-burners.com

The condensing boilers, with which I am most familiar, and recommend, are by:

AERCO International, Inc.
100 Oritani Drive, Blauvelt, NY 10913
Phone: 845-580-8000
Toll Free: 800-526-0288
http://www.aerco.com

The replacement burners my former company still recommends are made in Germany, by Weishaupt. You can find the representative for each state at: http://www.weishaupt-corp.com/mainUnternehmen/menuKontakt/menuKontaktWeltweit/central_america/USA/

Install New Technology Hardware

5. C - Improved hot water reset control

The merits of changing the temperature of heating water, as the outside temperature changes, should not be in question. However, there is ample opportunity to improve upon the traditional outdoor reset control strategy. This discussion outlines numerous areas of improvement. Now, I will tell you, up front, that I didn't figure out the features, advantages and benefits of improved hot water reset control. I learned about it as a manufacturer's rep for the Swedish manufacturer of temperature controls. I have just figured out a way to explain it in a way that I believe anyone can understand it.

The typical hot water reset controller, for commercial buildings and schools, at least in the upper Midwest, is most often set to maintain approximately 110°F water temperature, at 60°F outside air temperature. It then increases linearly, to as much as 200°F water temperature, at -20°F outside air temperature. Under this reset schedule, few people ever complain of being too cold, at least not due to a lack of hot water. Because cold complaints usually have some other cause, the reset schedule is rarely, if ever, looked at to see if it can be changed to a more efficient operation.

The proper reset schedule will depend on the type of heating system. The two main types of systems, considered here, are the radiation system and the air-handling system. For **existing** heating systems, a 15°F difference, ΔT, between supply water and return water temperatures, should be the goal for most fin tube radiation systems. Cast iron radiators and European style radiators should be able to have a 25°F ΔT. Systems with air-handling equipment should also strive for a 25°F ΔT.

However, for **new heating systems**, there are lots of advantages to designing to a 40°F ΔT. That means using more expensive, but more efficient heat transfer coils for the air-handling units. That can mean smaller pipes, smaller pumps and colder return water temperatures, which relates to more efficient condensing boiler operations.

These temperature differences should signal the building engineer that the building is utilizing the heating energy that is being generated and distributed. Lesser temperature differences indicate that an excessive amount of heating energy is being supplied and wasted. With the possible exception of early morning recovery from night setback, these efficient temperature differences are rarely found. There is good economy of operation to be gained by having larger temperature differences between supply and return water temperatures, especially if you operate with modern, efficient, condensing boilers.

The process of achieving increased temperature differences usually starts by lowering the 200°F, maximum water temperature, to 160°F. In 35 years in the industry, once simultaneous heating and cooling has been eliminated, I have seen only a handful of buildings ever needing more than 160°F heating water, even at the harshest of outside conditions. The exception is often apartment buildings, built with the cheapest available fin tube radiation, which may need 180°F, or more, heating water. If hotter water appears to be needed, first look for other possible reasons. In most cases, you'll find something broken or dirty. Sometimes, you may find something ill designed!

You saw the equation for heat loss in Chapter 3.D. Seeing it again may help in this discussion too.

Btu's of heat loss = square footage of heat loss area X U X Delta T (ΔT)

U = 1/R, R being the insulation rating of the walls, windows and roof

ΔT = the difference between the inside and outside temperatures

The calculation for heat loss from any building increases linearly, with colder outside temperatures. It's the only thing that changes in the equation. For that reason, the rate of increase for hot water reset schedules has always been linear. However, the Swedish findings, confirmed by my observations, have been that buildings are better capable of utilizing the internal heat gain from occupants, lighting, computers, and so on, as the outside temperatures drop. Therefore, proper comfort, and improved economy of operation, can be provided by a nonlinear outdoor reset schedule, as illustrated in Figure 5.C.1.

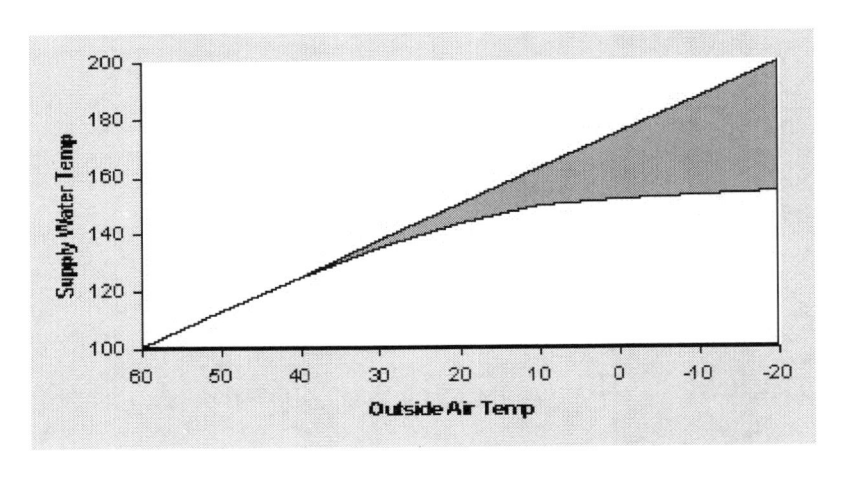

Figure 5.C.1 Illustration of savings from a non-linear outdoor air reset schedule.

Okay, that illustration of a nonlinear reset schedule is a little too accentuated. Raise the -20°F reset temperature to 160°F and you're there.

Referring now to the desired building temperature profile, for an intermittently occupied building, in Figure 5.C.2, line AB represents an occupied space temperature. Point B is the time of day that a perceptible drop in space temperature, for night setback, is allowed. Line CD represents the time during which the night setback temperature is maintained. Point D is the time at which the morning boost operation starts in order to achieve an occupied space temperature, at the time of occupancy, Point E. On a continuum, points F & A are the same.

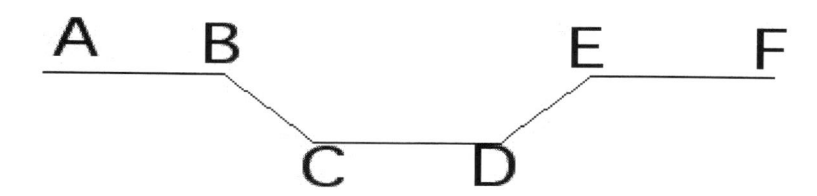

Figure 5.C.2 Profile of the desired day/night temperature profile, for an intermittently occupied facility.

The typical nonlinear outdoor reset schedule, must not only provide sufficient hot water to maintain the proper occupied space temperature conditions, E-B, but must also be capable of providing a relatively short morning boost period, D-E. Therefore, during the day, with the internal heat gain of occupants, lighting, computers and equipment, the typical reset schedule is way too hot. Most ΔT's will only be 3°F to 10°F, wasting lots of heating energy.

Water temperature that is too high, does not lend itself well to modulating controls. When the controls ask for a little heat, too much heat is provided, causing the system to immediately ask for no heat. Modulating temperature controls end up acting more like on/off controls, often causing comfort complaints. If controls do modulate, wire draw damage of valve plugs and seats, as described in Chapter 1.D, is certain to follow, because the valves are in the nearly closed position, most of the time. When the plug can no longer close against the seat, heating is continuous, as is the waste of energy. Comfort also suffers.

A modulating control valve operates best when it can throttle in a mostly open position. This requires an outdoor reset controller that can provide three separate hot water reset schedules. As illustrated in Figure 5.C.3, one reset schedule for occupied operation; a reduced water temperature schedule for unoccupied times; and finally, a highly inflated reset schedule, the likes of which most systems operate 24/7, for just 1 to 3 hours of morning boost.

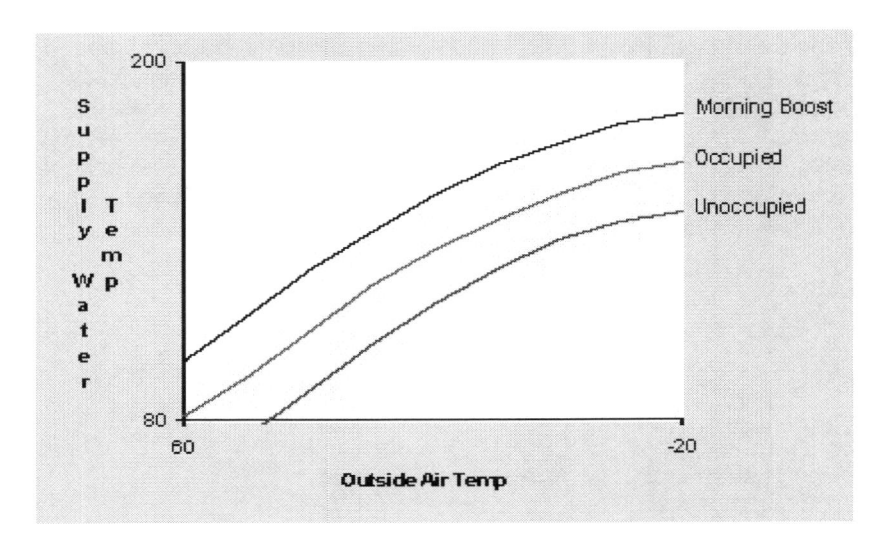

Figure 5.C.3 Illustration of three (3) separate hot water reset schedules.

Reviewing once again, the desired temperature profile shown in Figure 5.C.2, Point B is the time at which a perceptible drop in space temperature is okay. Given the flywheel effect of a building to maintain its occupied temperature, the time at which the controller can activate the unoccupied water temperature schedule can usually be well in advance of the time when the building is unoccupied. Similarly, the time at which the system goes into a morning boost operation, D, should also be optimized for maximum setback duration.

As described in Chapter 5.B, the purpose of these efforts to provide the coolest possible heating supply water, is to provide the coldest possible return water temperature to condensing boiler(s). When you buy a condensing boiler, check the boiler manufacturer's efficiency curves, like illustrated in Figure 5.B.3. These curves will generally show how the ratings increase significantly above a 120°F return water temperature. So, every hour that you can operate below those temps, it's like money in the bank!

In addition to an outside sensor, this control strategy should have a space sensor, located in the coldest area of the building, or multiple sensors, to provide space temperature averaging. This information should then be used to provide the necessary feedback to the BAS reset controller to make automatic adjustments of the occupied reset curves and optimization of the setback and morning boost times. The water temperature is then simply increased from the occupied reset value for only one to three hours during early morning boost. Similarly, the water temperature is reduced from the occupied reset value for more than half the total heating season hours, for night setback. A shift up and down from the occupied reset curve of 25°F to 30°F usually works pretty well.

Figure 5.C.4 illustrates the temperature duration chart for the Minneapolis/St. Paul area of Minnesota. You will note that there is approximately 6,000 hours per year, when the outside air temperature is less than 60°F. Since that is the outside temperature, below which most hot water boilers and circulating pumps are activated, we will use that as defining the heating season for Minneapolis/St. Paul.

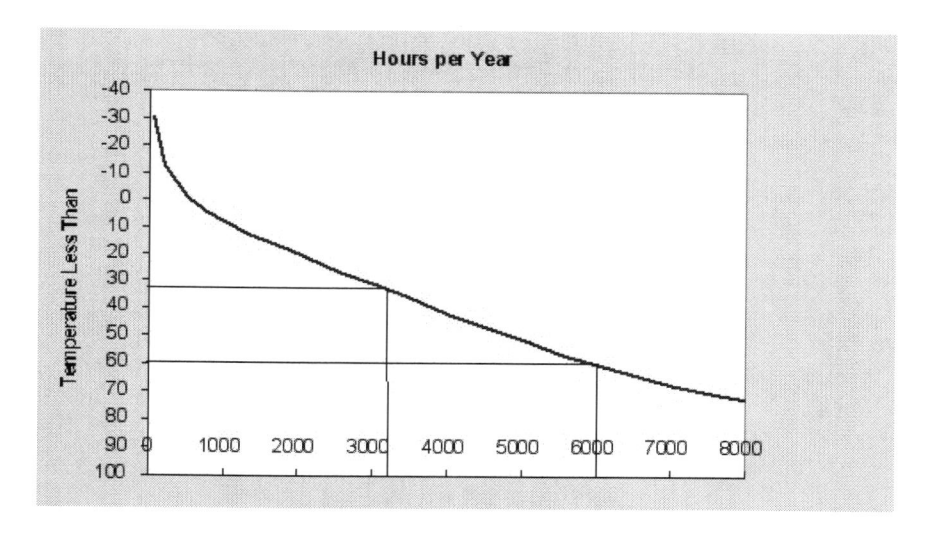

Figure 5.C.4 Minneapolis Temperature Duration Based on Three-Hour Observations.

Below 35°F outside air temperatures, there is no question that the hot water circulating pump(s) must run continuously for freeze protection. Above 35°F outside air temperatures, the hot water circulating pumps may or may not require continuous operation. With today's construction standards, if the outside air temperature is above 35°F, and all of the exhaust fans are properly turned off, the space temperature of nearly 100% of all commercial buildings and schools will not drop to the desired night setback temperature overnight. In fact, most buildings won't even lose that much heat over a weekend. Therefore, all of the heating and pump energy, normally expended between B & C, on Figure 5.C.2, when the outside temperature is above 35°F, is wasted!

In a typical office or school facility, unoccupied 60% of the time, improved hot water reset control could provide total heating system shutdown, for approximately 1,900 hours a year. That's 60% of the hours between 35°F and 60°F, in Minneapolis and St. Paul. It equals roughly 30% of the area's defined heating season! Remember, that includes pump energy too. You can use your area's temperature duration chart to help determine your savings.

Again referring to Figure 5.C.2, the desired occupied time is at E. As mentioned earlier, the BAS controller should have the sophistication to optimize the heating-water-system boost start time, D. If this hot water is being circulated to fan systems of any description, the fan systems should also begin staging on, at approximately the same time.

As mentioned earlier, the heating water system can go into a reduced setback condition, well in advance of the unoccupied time. However, if there are associated air-handling systems, you will want these units to continue to run for ventilation and possible atmospheric cooling, until the unoccupied time.

When using this controller on a steam to hot water converter system, switching back and forth between water temperatures causes no special problems. Today, however, most noncondensing hot water boiler manufacturers ask that you use a 160F minimum supply water setpont, to keep a minimum water temperature above 140°F, to avoid condensation problems. So, if your steam boiler is not needed during the times that the pump(s) are off, you can also turn off the boiler. Just make certain that it maintains a minimum water temperature of 160°F during these unoccupied times and the boilers have high fire lockout to keep from going into high fire, below water temps less than 190°F.

If you have hot water boilers, direct reset of the boiler water temperature can only be done with condensing boilers, specifically designed to do so. For noncondensing boilers to circulate less than 160°F heating water, a valve that mixes hot boiler water with colder return water, that is, a mixing valve, must be used. Most contractors, unfortunately, would use a 3-way mixing valve.

Another area of precaution is the potential for thermal shock problems that may arise in noncondensing boilers, especially when switching from the night setback water temperature to the morning boost water temperature. During this time, the return water can be as much as 60°F to 100°F colder than the boiler water temperature. If this water is allowed to quickly return to the boiler, with a fast operating 3-way mixing valve, thermal shock damage will likely occur. To alleviate this problem, a 4-way mixing valve should be used, as illustrated in Figure 5.C.5.

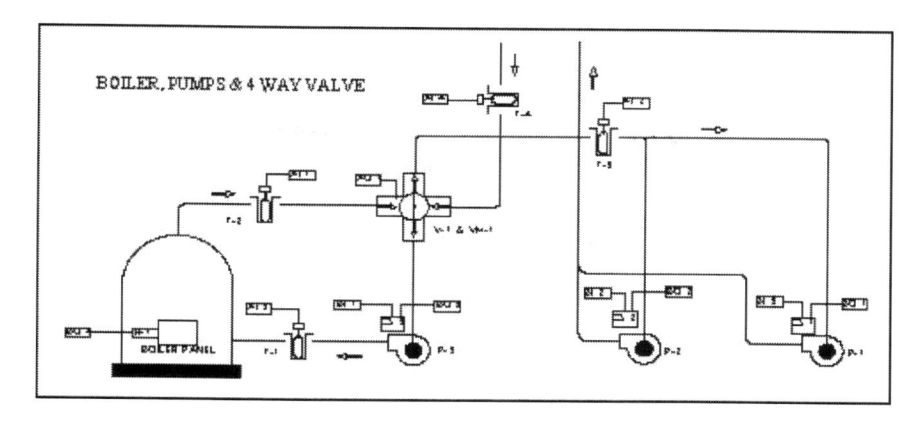

Figure 5.C.5 Four-way rotary control valve to protect the boiler against thermal shock

With the use of a 4-way mixing valve, improved hot water reset control can sense and maintain a minimum return water temperature to the boiler, no colder than 30°F less than the boiler water temperature. Again, this will provide a reset water temp less than 160°F and help guard against any thermal shock problems in noncondensing boilers.

For a moment, let's say that improved hot water reset control looks like something that you would want to incorporate into your facility. At this point, one needs to consider the age and condition of the existing boilers and burners. The installation of a 4-way mixing valve, with controls, can easily exceed $13,000. Then, there is the issue of finding a control contractor that fully understands the concept of this discussion. Back in the 80's, the controller that did this was off-the-shelf technology from a Swedish manufacturer, Tour and Andersson. From what I can find, they now operate under the name TAC (www.tac.com), which is owned by Schneider General, but I'm not getting anyone there to return my e-mail inquiry. Since few companies know how to provide this unique control strategy, you may want to share this with your BAS contractor.

With noncondensing boilers, condensation will generally occur somewhere between 83% and 88% efficiency. By the time the heating season is over, the constant on/off operation of most oversized, noncondensing boilers, that efficiency will drop significantly. I encourage you to ask whether or not it is wise to make that kind of investment in aging boilers and burners that have, at best, a seasonal efficiency of about 70%? Even without the state-of-the-art hot water reset control, outlined here, just going from a 70% seasonal efficiency, to a 92% seasonal boiler efficiency, you will reduce the annual heating energy consumption by nearly 24%!

As a contractor, I was completely sold on condensing boilers. Because of my understanding of this control strategy, I was able to have our BAS department incorporate it into the normal course of their work.

For use on the two hot water district heating systems on which I worked, the return water temperature, back to the district, must **not** be greater than 160°F. District heating systems typically generate 250°F, high pressure hot water. To maintain high efficiency, district utilities want to make certain that if they are going to send you heat that you use it. If you don't, they will penalize you! This means the end user facilities need high efficiency frame and plate heat exchangers and the heat transfer coils, at the air-handling units, also need to be designed for maximum heat exchange. A 40°F ΔT would be great. Your BAS contractor can just as easily provide a high limit to the return water temperature. If comfort problems exist, system inspection will usually show you why. Don't just automatically assume that the heating supply water temperature isn't hot enough.

In retrofit, or new construction, this type of hot water reset can be used to augment, and optimize, other night setback features. However, there are certain facilities that are not provided with adequate night setback capabilities. In these systems, improved hot water reset could provide the only method of night setback. As an example, many office buildings are constructed with perimeter radiation and variable volume air-handling systems. In significant numbers of these buildings, self-contained radiator valves are controlling the perimeter radiation. In others, a single temperature thermostat controls both the radiation and the variable volume box, in sequence. With both of these systems, as soon as the temperature drops just a couple of degrees, these heating valves are wide open. Simply turning off the air-handling system does nothing to provide the ASHRAE required night setback of the heating system. The above-described outdoor reset control would fulfill that requirement and provide the desired economy of operation.

In an effort to help protect against freezing coils, many air-handling unit control valves are driven fully open at night. Because of the internal heat gain in most buildings, and the excessively hot heating water, heating water flow is practically nonexistent for much of the day. I have personally been standing in the hallway of a school when the system went into night setback. Within minutes, I could hear the heating pipes in the ceiling popping and creaking from the expansion of having full flow of hot water for the first time in hours. Shortly afterwards, the boiler started and ran nonstop for hours! My servicemen have even found systems in which the excessive heat has melted fire damper linkages and set off fire alarms.

Figure 5.C.6 Babbling brooks do not freeze!

Circulation of cooler water, from improved hot water reset control, will reduce energy consumption and help keep the coils from freezing, when used properly.

How much energy will this control strategy save? The State Energy Office, SEO, of the Minnesota Department of Commerce, allowed contractors to use a 20% annual reduction of heating system consumption in the payback calculations. I had provided the SEO with sufficient documentation that they did not require any additional calculations. Some of that documentation includes the fact that this control strategy was off-the-shelf technology, throughout Europe, over 40 years ago.

Install New Technology Hardware

5. D Building automation systems - BAS

When setting out to purchase the installation of a BAS, I believe the most critical issue is not the equipment. It's the skill, knowledge and commitment of the installing contractor that is most important. If you've been reluctant to use Best Value procurement on other projects, here is where you really need to make an effort to use it. Take the following discussion to the bank!

The main problem with temperature control systems is that things change. Ventilation rates change; improved sequences of operation are discovered; classroom and office functions change; and so on. Each time this happens with a pneumatic temperature control system, the panel has to be reworked. This can often be done by simply re-utilizing the existing controls. But, as you can imagine, just by looking at the pneumatic temperature control panel below, it takes considerable skill to do so.

Figure 5.D.1 Pneumatic temperature control panel.

Now, I'm not saying that it doesn't take considerable skill to make a BAS do what you want it to do. But, I believe one of the keys to my company's success was in asking our pneumatic temperature control specialists to continue their progression of learning to the next level. As pipefitters, experienced in all phases of HVAC, it just seemed natural that they be the ones to install, and commission, electronic BAS. Some parts are really not all that complicated. The output signal from a pneumatic controller is 3 to 13 psi air pressure. It exactly corresponds to the 2 to 10 vdc output signal from a BAS controller. What takes extra skill is in knowing how to do the programming, in both systems. The advantage is to have BAS department personnel know what the mechanical equipment should and should not do in response to that control signal. When system requirements change, with a BAS, all the work is done from a keyboard. And, with the help of the Internet, monitoring and changes can be made from anywhere in the world.

Boiler rooms can often be the best place to start the process of adding a BAS, to older facilities. The improved hot water reset control of Chapter 5.C demonstrates just some of its benefits. Converting antiquated steam heating systems to hot water heating; replacing old, oversized boilers; controlling new boilers with multiple temperature hot water reset control strategy; and upgrading temperature control systems to eliminate simultaneous cooling and heating, can possibly bring savings of as much as 50%! The technology is available. In time, converting to a BAS for air-handling unit control may be your only option.

As I keep repeating, the key to any improvements involving contractors is to find those that you know, like and trust. I'm convinced that there are contractors in your area that will work well with the Best Value process described in Chapter 5.A. This industry is still changing. Search your industry's professional directories for the contractors that you need. Do their installers have a history in pneumatics? Do they know mechanical systems? Ask them to dial up an existing system from their laptop. Your system should be able to do that too. If they can't do it in a remote demonstration, thank them for their time, but don't ask them for a Best Value proposal. I doubt that they qualify to provide it. Question them about simultaneous cooling and heating. Compare what they say to what you've learned here. Ask them to describe their method of hot water reset control. Remember, initially, you're just narrowing the field to find those you want to make a Best Value proposal. The Best Value proposal is not going to come from the contractor that has the best equipment. It's going to come from the contractor that best understands your systems. As stated in the ASU presentation, the Best Value contractor will usually present themselves. Hire them!

You may have noted that, up to this point, I have refrained from mentioning my former company's name. If I'm going to tell you about products that work, it's only fair that I mention them too.

Climate Makers, Inc.
1700 Freeway Blvd. Suite 10
Brooklyn Center, MN 55430
763-786-5999
www.climatemakersinc.com

Northern MN Office
11075 Thiesse Road
Brainerd, MN 56401
218-825-0145

Install new technology hardware

5.E - Variable Frequency Drives (VFD's)

Once again, I'm old enough to have been in the industry before variable frequency drives were even invented. When they first came out, there was a great deal of skepticism. They were new and many people just didn't trust them. VFD's came out shortly after the concept of variable air volume (VAV) air-handling units came into vogue. That means that, in the past 30 years, there were hundreds of thousands of VAV air-handling systems installed with alternative methods of volume control, mostly inlet vane or discharge air dampers. I'm assuming that there have to be a ton of these still in operation. Let me explain why these units should be retrofit with VFD's.

At 100% airflow, a motor consumes 100% of its designed energy. When providing VAV control with dampers, the best you can expect in electrical consumption is a nonlinear reduction to approximately 75%, even at zero airflow. Variable frequency drives operate on a cube root basis. That means that at 90% airflow, the electrical consumption is 0.9 X 0.9 X 0.9, or 72.9% of maximum. So, at 90% airflow, VFD's are already more efficient than with damper control, at any volume!

Most VAV air-handling systems try to average in the range of 60% air flow. That's 0.6 X 0.6 X 0.6, or 21.6% electrical consumption, compared to about 75%, at its best, with damper control. Bottom line, if you've got any VAV air-handling systems, with damper volume control, you really need to look at VFD control. Here is the calculation formula to determine electric motor operating costs:

hp X 0.75 kW/hp X hours of operation/month X $ (electric rate/kWh)

This will give you the cost of operating the motor without any variable volume control. With VAV damper control, take that value times 0.75. Then take that cost times your best estimate of percentage of average airflow, in decimal form, as in previous examples. Now take the cost of a VFD installation and divide by the projected monthly savings. That will give you the return on investment, ROI, in months.

On this topic, I'd like to share the following story. Water pumps equipped with VFD's save energy the same way that air-handling units do. Swimming pool circulation pumps are another great application for VFD's. Swimming pools are required to operate at four water exchanges per day. As the filter(s) get more and more plugged with debris, the water flow varies. To maintain a relatively constant flow, pool operators are usually required to manually adjust 90° turn ball valves. In terms of savings, this is very similar to volume damper control on air-handling units.

In a real life situation, I once found a 25 hp pool pump, which probably should have been only a 15 hp pump, maybe even 10 hp. It was consistently squeezed down by 50% to over 70% of its maximum flow capacity. To solve the problem, I proposed a VFD installation. It turned out that one of the school board members was a union pipefitter. Believing that my company was a non-union contractor, he did everything possible to argue against my company doing the work. Finally, he convinced the board that they could do the work themselves.

VFD's are voltage specific. This installation had 240 volt, 3-phase power. The school's VFD provider didn't have a 240 volt VFD, so knowing that the voltage actually received is plus or minus 10% and that the acceptable voltage of the VFD's are plus or minus 10%, the supplier felt comfortable in selling the school a 208 volt VFD. When installed, however, the VFD overheated terribly. By the time I saw the installation, they had a floor fan tilted up, blowing a constant flow of air on the VFD to cool it. It lasted slightly longer than the warranty, before having to be replaced with one of the proper voltage!

To finish the story, some years later, one of our servicemen was teaching a class at the vo-tech school used for pipefitter training. The above mentioned pipefitter/school board member was also teaching a class that night. Knowing our serviceman by name, but not by company, he saw our truck in the parking lot and asked our serviceman what that ^%*$#! non-union truck was doing in the lot. He was shocked, and embarrassed, to finally learn it was our union pipefitter serviceman's truck.

The moral of the story is, put your faith in people with whom you have come to know, like and trust. Make certain that your supervisors understand why you know, like and trust the contractors of your choice, union or non-union. Choosing the right contractor isn't just about which contractor has the lowest price. This subject is discussed more in Chapter 5.A, Best Value and the bonus Chapter 7.A on Best Practices.

Install New Technology Hardware

5.F - Summer swimming pool heaters

If you don't have an indoor swimming pool; never plan to have one; or never plan to work in a facility that has one, you can simply skip this chapter. Just know there is valuable information here, just in case.

If you're heating your pool in the summer, with the building's main boiler plant, you're wasting a ton of money. More importantly, perhaps, you're adding years of unnecessary wear and tear on the boiler plant. I gave you some operating tips in Chapter 3.A. Recheck those tips to save enough money to buy one, or more, small summertime pool heaters. When funds are available, here are some guidelines.

Let's make certain that you're buying a swimming pool heater that's not any bigger than it needs to be. In my experience, swimming pools are drained infrequently. Perhaps, only once every 4 years, or so. When the pool is refilled is the only time that you need lots of heating capacity. If you size the water heater for that capacity, it will, more than likely, not operate in its most efficient mode the rest of the time. It will likely short-cycle. Swimming pool water heaters should be sized for the steady state heat loss of an inactive, uncovered pool. With the proper use of a pool cover, the heater should be capable of recovering from the day's activities, before the cover is removed for the next day.

The chlorine in swimming pools makes the water very corrosive. Therefore, there should be a stainless steel, frame and plate style, heat exchanger between the pool heater and the pool water. Control of the converter water temperature, should be a direct reset of the summer boiler, based on the converter return water temperature. The discharge temperature must be warm enough to heat the pool, without being hot enough to scald, so a high limit on the discharge temperature would be in order. Even with a modest 5:1 turndown ratio on a not too large boiler, you should be able to have a nearly steady, efficient runtime on the pool water heating system. Again, condensing boilers have much higher turndown ratios. Two small condensing boilers, operating at part load, would make your system even more efficient.

To recover from the ground water temperature, once every few years, when the pool is refilled, you've already got a system that is capable of heating the pool water. The new, smaller system should be in parallel with the current system. If the old system is not a large frame and plate heat exchanger, and something goes wrong with it, install one that is properly sized. Don't waste money on any controls for the once-every-four-years system. When you're close to the desired 80°F pool water temperature, valve off the building heating boiler and maintain the pool water temperature with the small, condensing style, pool heating boiler(s).

Install new technology hardware

5.G Swimming Pool Blankets

On more than one occasion in this book, I have mentioned the use of pool blankets. Most of the excuses for not having a pool blanket center around the tasks of putting it on and taking it off. No one wants to do it. For some years, installation and removal has been as simple as turning a key. With the excuses removed, and the proper control sequences clearly described in chapter 4.C, I cannot encourage you more to get a pool blanket for your indoor swimming pool.

Figure 5.G.1 Automatic pool blanket.

At one time, there were two primary automatic pool blanket manufacturers selling pool blankets in my area. One is listed here. The other was called Löf something, but I'm not finding them on the web. Pool blankets work!
Alta Enterprises, Inc.
www.altaenterprises.com
(800) 624-1235
poolcovers@altaenterprises.com

Install New Technology Hardware

5.H Lighting

If you haven't done any lighting retrofit projects yet, now is the time to get something started. I'm not a lighting retrofit specialist, but even before I retired, I saw some pretty neat things happening in this field. Compact fluorescent bulbs can replace any incandescent bulbs. In fact, most incandescent bulbs can no longer even be manufactured. Now, LED's seem to be all the rage.

So, find a lighting specialist that you know, like and trust. They'll be able to tell you about what's new; help you know for how long you should plan to keep bulbs off before just leaving them on; and help you determine the ROI.

My only other contribution about lighting regards their sheer numbers. Don't just look to replace an old bulb or fixture with a new one. Look at the desired lighting levels. I frequently still see large areas of hallways considerably over lit. My belief is that they should be lit for safe ingress and egress. Many are too bright even for task lighting! Reduce the number of fixtures to just what's needed.

Preventive Maintenance

6.A – Training

At my business, each new employee would spend most of their first day with me. In the preface, I mentioned taking lots of pictures and making training presentations to lots of different groups. Some of those photos have been included in this book. I used the same presentation to introduce our new mechanics to our way of doing things. Around the office, it was nick-named "The SSDD Presentation". SSDD stood for "same s - - t, different day", because of the constant occurrence of finding the same items broken and dirty, over and over again.

With this presentation, from day one, my new mechanics understood that they were partners with our clients' in-house personnel. They also understood that, despite their considerable skills that led them to us, there was a good likelihood that, from time to time, they were going to get dirty. Each employee would be expected to solve the problems presented to them. If something was broken, they needed to fix it. If it was dirty, they needed to clean, with or without the help of the in-house personnel. As one of our servicemen frequently said, when he found himself in the middle of a not-so-pleasant task, "It all pays the same!" It's a good attitude to instill in your employees.

By now, you've probably got a good idea of what this book can do for your maintenance department. If you believe, as I do, that it can have a positive impact; I'd encourage you to buy a copy for each of your maintenance personnel. It's a good idea to have everyone in your department on the same page. It's also a great idea to have management and board members familiar with your maintenance plan. It's priced so that everyone can easily have their own copy.

Throughout Minnesota, many gas and electric utilities provide periodic training. Check to see if your utilities provide any training programs. There is also a national training program called the Building Operator Certification (BOC) program, often endorsed by utilities. www.theboc.info

The Minnesota Association of School Maintenance Supervisors, MASMS has numerous levels of certification through the training program for their members. If your area has a similar statewide organization, it would be worth checking with them. www.masms.org

BOMA, www.boma.org and the Association for Facilities Engineering, AFE, www.afe.org also have facility engineer training programs.

Later, when I started e-mailing my page-at-a-time book, our servicemen received them too. It absolutely made them better servicemen. It also helped reinforce the integral role they play in developing the know, like and trust relationship we wanted with each and every client. The only excuse for our servicemen not sharing their knowledge with in-house personnel, was not knowing what they were doing sufficiently well, or being reluctant to ask for help. With a super knowledgeable service foreman, and a number of other experienced servicemen, we always had someone, with whom they could consult. If needed, our service manager would join them on the job. I also encouraged our servicemen to take additional union provided training, as well as providing our own periodic training.

Short of hiring well-trained union pipefitters, like I was able to do, as an HVAC contractor, hiring vocational school graduates might be a good idea. But, like I've mentioned frequently, I encourage you to work with the service contractors that you know, like and trust. Let your contractors know that you plan to have one of your employees shadow the serviceperson that they send to work in your facility. At the very least, that employee can be a good assistant, often times, shortening the length of time the serviceperson is in your facility. In time, you can expect your in-house personnel to be able to perform more preventive maintenance; recognize problems earlier; and even fix many of the problems found.

In 1973, during the oil embargo that started the energy crisis, I was introduced to a booklet printed by the Educational Facilities Laboratory entitled, **The Economy of Energy Conservation In Educational Facilities**. The opening statement of the chapter on maintenance and operations stated:

"Energy waste springs from two basic sources, lethargy and ignorance."

For nearly 35 years, I repeated that quote every working week - often, more than once a week! My family even tired of me repeating it, when not at work! Consider this book my contribution to combating the ignorance (lack of education). Just by reading this book, you've already demonstrated that lethargy (laziness) is not an issue!

Preventive Maintenance

6.B - Planning

Facilities need, at least, a 1-year and 5-year facility plan. In their Best Practices Review, the Office of the Legislative Auditor for the State of Minnesota, suggests a four-step process:

1. Plan
2. Prioritize
3. Budget
4. Implement

Planning should start when you first buy, or build, your facility. I believe you should be able to look at a commercial/institutional building, at any point in its life, and see a facility that has a 40 year life expectancy. At least, that should be your stated, and published, goal. That means, within a few years of its construction, the building owner should start budgeting, and reinvesting, 2.5% of the facility's value, every year. That's 100% of the facility's value, divided by 40 years, equaling 2.5% per year. It's a simple formula, but one to which too many people do not adhere. The result is an inability to budget 25%, or more, of the facility's value, when things really start going downhill. In those first years, I would suggest putting 0.5 % to 2.0% of the building's value into a savings account for future major projects.

That said, start by documenting current facility issues. Even if you can take the management team on a tour of your facilities, take pictures, building a portfolio with all available information. Put the information in a document that you can e-mail to all of the decision makers. The more people that know your facilities have issues, the better. I can't tell you the number of people that are reluctant to do that, because they've been the one in charge for the past 20 years and their building isn't in very good shape. You're learning new things all the time and this is just the latest.

Demonstrate your new skills. You'll have your facilities' benchmark analysis from Chapter 6.D to help demonstrate the need. Your facilities need you to constantly grow. Documenting the issues helps put the ball in someone else's court, not yours. You're just working with what they give you! Keep your plan updated. Demonstrate that you need more, when you need it. It should not be a stagnant document.

Unless you have limitless resources, you'll need to prioritize each item. One director of buildings and grounds color codes his plans on a spreadsheet. Virtually every item on the 1-year plan is coded red. The 5-year plans are coded as follows:

Red - This item requires immediate attention. This item needs to be finished in the next 1-year plan. If we do not correct this situation, health or safety will likely be jeopardized, or additional, costly damage may result.

Yellow - This item has cause for concern. If additional funds are available, we need to take care of this issue, in a timely manner.

Green - This item is on the radar screen, but is not causing an immediate health, safety or future damage issues.

When you first start, prioritization may even be needed within the maze of top priority red items. So, rank your needs to the best of your ability. Ask others in your organization for their help. Ask administrators to help you prioritize. Don't just take on all of the responsibility yourself. Spread it around. There is plenty to share. Remember, documentation is your friend!

At some time in the past, then superintendent of the Elk River, Minnesota Public Schools, Alan Jensen, made a presentation to one of the school official organizations, to which I belonged. In that presentation, Alan stated, "We're too poor to build cheap." Later he stated, "We're also too poor to not take care of what we have." At that time, The Minnesota Department of Education admitted to a $4 billion deferred maintenance deficit. In that same time frame, a survey of America's public schools admitted to a $112 billion deficit. Knowing that schools don't like reporting that things are as bad as they really are, I would suggest that the $4 billion Minnesota number was probably closer to $6 billion. If all 50 states just averaged what Minnesota was willing to admit, that would make the total US deferred maintenance number closer to $200 billion. That was back about 2000. If I were to guess, the dollar figure of deferred maintenance has probably continued to grow.

When it comes to figuring out when you'll need to consider major expenditures for equipment replacement, start with the following figure:

"Bathtub" Curve of Product Reliability

Product "burn-in" (Warranty Period)

Obsolescence (Replacement Period)

Useful life (Maintenance Period)

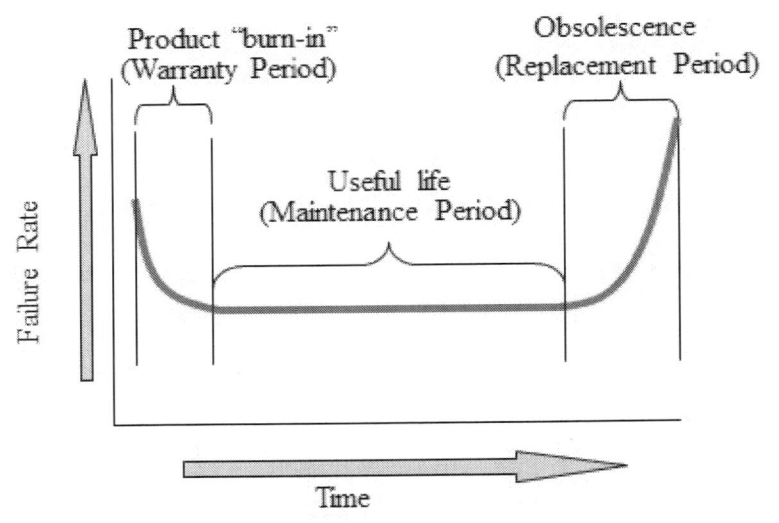

Failure Rate

Time

Figure 6.B.1 Bathtub curve of product reliability

Then add the following info to your understanding:

Useful Life Examples

- Lighting conduit and wires — 20 years
- Lighting fixtures — 15 years
- Air-conditioners — 15 years
- Boilers — 20 years
- Steam piping — 25 years
- Iron water pipe — 30 years
- Concrete sidewalks — 20 years
- Roofs — 20 years

Excerpts from *Means Facilities Maintenance Standards* Copyright R.S. Means Co., Inc.

Now, don't shoot the messenger. Some of these seem pretty short, but you'll note that I didn't generate the list. I'm just helping you see how short a time you have to tuck some extra money away. Once the big expenses start coming, the rest follow pretty quickly. Your plan should have you financially ready for them!

If you don't have any ideas of the costs of a needed project, consult the contractors that you know, like and trust. By now, I hope that recurring theme is really sinking in. If you want good service from a contractor, they need to be profitable. By having a good working relationship, you're going to know that they value your business and won't take advantage of you. However, that's a two-way street. Don't you take advantage of them either. They can't afford to continually invest their time, without being awarded the work. Every time you request quotes from a number of contractors, those that were unsuccessful have just added another cost to their overhead column. If you know, like and trust the needed specialty contractor, negotiate a fair price for the work and give them the contract. You'll save time. You'll get a better quality job. Everyone wins!

Many facilities operate under unrealistic financial constraints. I know of facility managers that are required to get bids for anything over $1,000. Today, no quality contractor can afford to respond to such requests. In Minnesota, even back in 2006, the legislature allowed public schools to negotiate projects up to $35,000 or $50,000, depending on the size of the community. By now, these values may even be higher. As you saw in the Chapter 5.A discussion, Minnesota also allows Best Value procurement. I don't believe Best Value procurements have any dollar limits – high or low!

Again, if you know, like and trust your contractor, these are realistic limits, within which everyone can live for direct negotiations. I don't believe anyone is out to take advantage of you. They know your relationship won't last if they do. If you want quality work, you have to give them an opportunity to hire the best qualified personnel and make an honest living. If you don't have a list of contractors that you trust up to those limits, you should be looking for new contractors. The first place to look is in your professional directories. The vendor members are those that support your organizations. In return, they deserve your support!

Preventive maintenance

6.C Scheduling

Gee, does anyone ever have the time to perform work on a schedule? Does anyone have the manpower to schedule preventive maintenance? Not many, that's for certain. The best illustration I have, in regard to scheduling, is the following story on changing filters in a school.

About the middle of November, Mrs. Johnson gets cold and calls the maintenance department. After first checking a few items, not the least of which is the thermostat, with no improvement, it is finally discovered that the air filter on the unit ventilator, or air-handling unit, is plugged. With a special trip to install a clean filter, comfort is restored to the classroom within minutes. Problem solved!

Later that day, however, Mrs. Johnson has lunch with Mr. Schwartz, who also mentions that his room has been cold. He also calls the maintenance department to register his comfort complaint. After first going through a similar list of potential problems, and another special trip for a clean filter, the air filter serving Mr. Schwartz's classroom is also changed. Problem solved - again!

Over the course of the next two weeks, this scenario is repeated again and again, until virtually all filters have been changed. So, when Christmas vacation arrives, changing air filters is no longer on the schedule. But, by mid to late January, it starts all over again. So, here is a task that, if done on schedule, it could be done in a matter of a few hours. Proper timing for changing filters in a school is: before November 1st; during Christmas and spring break vacation times; and again in the summer, But, when done one at a time, over the course of several weeks, the hours add up to days, not to mention the interruption to other daily maintenance tasks.

Here is another example of not performing preventive maintenance in a timely fashion:

Figure 6.C.1 Priceless photo.

In the above photo, numerous ceiling tiles have collapsed from the weight of the water that has leaked through the roof. In that same classroom, an active light troffer had water pooled on the plastic light diffuser. Waste baskets are scattered about the classroom, in an effort to collect as much of the rain as possible. In addition to the obvious, all this water can also lead to a number of indoor air quality issues. You can apply the six-step program to more than HVAC.

Scheduling timely roof repairs; boiler upgrades; the conversion from steam to hot water heat; updating from pneumatic temperature controls, to BAS; and the replacement of other large ticket items, all expand on the principle of scheduled preventive maintenance. A one-year and five-year plan to get these items on the calendar is essential.

Preventive maintenance

6.D Benchmark Analysis (Reporting)

There is a great tool that I encourage all facilities to use, at least, once a year, or monthly, if you can. Okay, **I believe** it's a great tool, because I designed it. I find that most facilities don't know their heating energy consumption rate. It may be too difficult for them to calculate, or if they do calculate it, they don't know how to interpret it. The heating energy consumption rate is a simple calculation of Btu/square foot/heating degree-day. This "benchmark" analysis gives owners of multiple buildings the ability to compare facilities to each other; to themselves, from one year to the next; and to other facilities throughout the region or industry.

Why is this so useful? First, and foremost, it lets your administration know that you're doing a good job. If you don't tell them, they won't know. By providing an ongoing history of all facilities' energy consumption, it really tells a story. If you replaced a boiler burner last year, this will help illustrate the economic justification of your good recommendation. If the new burner didn't help, it may give you ammunition to go back to your contractor and ask why. Better to know that in year one, than year 15!

Even if your numbers aren't all that good, when you first report your findings, it demonstrates that you're working on a plan to make improvements. You can't make a plan to go somewhere, if you don't know where you are. This is your starting point.

If the administration complains about busting the energy budget, the annual benchmark analysis helps you illustrate that it was the number of heating degree-days, or increased cost per unit of fuel, that caused the problem, not the rate of consumption. I believe, if you followed the first 5 steps, you can't help but lower your energy costs.

This calculation is important to understand the basic need for energy conservation. Even with increased ventilation requirements, a steam heated facility should be able to operate in the range of 6.0 Btu/square foot/heating degree-day. A really well operated steam heated facility may be able to operate as low as 5.5 Btu/square foot/heating degree-day. If you're operating at much lower than that, you may have a need for investigating the ventilation rates. You may not be providing adequate outside air for your occupants. If your consumption is significantly higher than 6.0 Btu/square foot/heating degree-day, you definitely have good economic justification for starting an energy optimization program, like outlined here.

Hot water heated facilities are much more economical to operate. I have hot water heated schools that consistently operate in the range of 3.5 Btu/square foot/heating degree-day. This is a figure, with which at least one person with the Wisconsin Focus on Energy program and I agree is an appropriate target for most hot water heated schools. Commercial buildings, with less internal heat from occupants to help with the task of heating, may have to use slightly higher targets. No matter what the heating system design, natatoriums (indoor swimming pools) add between about 0.5 and 1.0 Btu/square foot/heating degree-day to their target cost of operation.

At one time, benchmarking was a complicated, lengthy discussion on Btu content of fuel, heating degree-days and so on. No wonder no one wanted to benchmark their facilities. So, I finally made my own spreadsheet. All you have to do is enter the various heating energy consumptions; associated costs; and heating degree-days, all from your utility bills. The benchmark spreadsheet will convert all fuel types to Btu's for you. You can do this on an annual basis, or better yet, enter the data monthly.

A real, but abbreviated, benchmark analysis for a district with all hot water heated facilities, looked like this:

Benchmark Analysis

2004/2005

School	Btu's/sq. ft./HDD	Target	Savings
Elementary	8.7	4.0	$9,200
Middle	7.6	4.0	$13,400
Jr. High	6.8	4.0	$12,700
Senior High	9.3	5.0	$21,700
		Total	$57,000

I believe you can see that, as this table grows from month to month, or year to year, how useful a tool it would be. With the benchmark spreadsheet, you'll be able to easily do this yourself. Also, when it comes time to replace your old boilers, this benchmark analysis can help you properly size and project the savings from your new boiler system.

I've found that most heating and electrical utility bills will provide the local heating and cooling degree-day information for your specific area. If your utilities don't provide this, here is where you can get it for 14,000 weather stations in the United States and Canada:

http://www.weatherdatadepot.com

Note that this is a heating benchmark analysis, but you could use it for an analysis of electrical consumption. Done just for the summer months, it gives you an opportunity to analyze your air-conditioning costs, using cooling degree day data.

To download your copy of the benchmark analysis, complete with instructions, go to: www.sixstephvac.com

Preventive maintenance

6.E - Commissioning

At the March, 2008 meeting of the Minnesota Chapter of ASHRAE, Dick Pearson, ASHRAE Fellow and principal of Pearson Engineering, in Madison, Wisconsin, made a presentation to the Minnesota Chapter, ASHRAE membership. His purpose was to inform the members that ASHRAE was endorsing the 30% energy saving challenge that BOMA announced in 2007. I first mentioned that challenge in the preface.

Dick sighted a couple of example facilities, for which all they needed to do, was to start paying more attention to what was happening in their facilities. In these instances, the building automation system had data logging capabilities, but Dick suggested that similar results are available with other investigative tools as well. Once the maintenance manager took time to study the data, not just look at it, operational improvement opportunities jumped off the page. Now, it took an energy specialist to convince management that the savings were there, and worth the effort to look for them.

I've been telling people of these opportunities, in one way or another, since 1973. My company confirmed these savings over and over in the early 80's, when schools were required to have energy audits. Savings of 20% in all commercial facilities were confirmed by the Herzog/Wheeler study, performed for the University of Minnesota, in 1992. And still, all these years later, BOMA and ASHRAE state that there is still a 30% energy savings possible in your facilities.

A new industry, called commissioning, has actually been created to specifically address making your building(s) operate the way they should. It's really all that this book represents. I just haven't tried to rename it to make it sound more current. It's just a lot of hard, and often times, dirty work.

How do you commission a building? While I don't know anyone else that states it exactly the way I do, it's done by:

1. Fixing what's broken
2. Cleaning what's dirty
3. Changing operations made possible by steps 1 & 2
4. Making temperature control system revisions to eliminate simultaneous cooling and heating
5. Installing new technology hardware
6. Preventive maintenance

How do you commission a building most economically? I believe no one can provide the majority of these services more cost-effectively than well-trained, in-house personnel. Combined with competent HVAC service and control contractors, your facility can operate the way it's supposed to, without the additional cost of having the building "commissioned".

Preventive maintenance

6.F - Do it!

The last item on the outline is simply called, "Do It", referring to getting out there and making it happen. I must have originally written this, when Nike had their "Just do it!" advertising campaign. Make what happen?

1. Fix what's broken

2. Clean what's dirty

3. Change methods of operations made possible by steps 1 & 2

4. Revise temperature control sequences, to eliminate simultaneous cooling and heating

5. Install new technology hardware

6. Preventive maintenance

Okay, I think I've emphasized the six-step process enough!

So, we have now come full circle. You now have many more tools, with which to better manage your facilities. Just always remember, 75% of maintaining proper HVAC systems, is simply keeping systems clean, dry and lubricated. No one can provide those services more cost-effectively than well-trained, in-house personnel. Don't be afraid to invest in your maintenance personnel.

BONUS - More things that are good to know, but don't fit into the six steps

7.A - Best practices

A recurring topic with school district, government and private buildings has been the frustration about the lack of input from the maintenance personnel in the design process. This could be in the design of a new facility or the design of an addition for an existing building. We all know how valuable the knowledge and opinions are of the people actually working with the equipment in a facility. Their views should be heard. Well, they have. The State of Minnesota Legislative Auditor's Office has released a Best Practices study for local governmental buildings. This includes some very interesting information about the proper operation of the types of buildings in which most of you work. The report lists what is working and gives concrete examples. I believe you will see many similarities with what you're finding in this book.

I encourage you to read, at least, the single-page summary to see where you stand. If you're having difficulty selling the concepts of this book, perhaps, this summary can be helpful in backing up the plans that you'd like to implement, showing how others have succeeded.

The summary can be viewed and printed at:

http://www.auditor.leg.state.mn.us/ped/bp/0006sum.htm

The full Best Practices Report can be found at:

http://www.auditor.leg.state.mn.us/ped/bp/pe0006.htm

There are versions for Adobe Acrobat, or plain text, depending on the software you may have on your computer.

If necessary, you can also obtain this information from the Office of the Legislative Auditor. Their address is:

Office of the Legislative Auditor
Room 140
658 Cedar St.
St. Paul, MN 55155-1603
Phone: (651) 296-4708
Fax: (651) 296-4712
E-Mail: Legislative.Auditor@State.mn.us
Hours: 8:00 A.M. - 4:30 P.M.

BONUS - More things that are good to know, but don't fit into the six steps

7.B Energy saving performance contracting

The basic idea is that a contractor performs specific energy saving services, and if the work doesn't provide the savings, as promised, the contractor has to pay the facility owner for the difference between what was promised and what was delivered.

From my experience, energy saving performance contracting is primarily for public schools, but it certainly can apply to other public and private facilities. In Minnesota and Wisconsin, special legislation has been passed to allow public schools to enter into performance contracts, for as long as 15 years. And, from what I'm reading, it is probably allowed in Minnesota government buildings, as well. I'm certain other states also have similar legislation.

Now, some of the following discussion comes from a career of selling against energy saving performance contractors. Mostly because small companies, such as mine, can't afford to finance millions of dollars in projects. If you choose to work with someone on a performance contracting basis, that's your choice. I just want it to be an informed decision.

The opportunity for performance contracting is entirely in the hands of the owner. If much of what is described in Chapter 1, Fix what's broken; Chapter 2, Clean what's dirty; Chapter 3, Changes in operations; and Chapter 4, Revise temperature control sequences, still are not done, your facility is what many performance contractors call "Low hanging fruit"! Your buildings are exactly what they're looking for.

If you didn't get it the first time through, let me repeat. Done by my company, rarely did all the work in chapters 1, 2, 3 & 4 ever have a return on investment (ROI) that exceeded one year. Now, labor and material costs have gone up and the cost of natural gas has gone done. So, double or even triple the ROI. That's still 50% or 33% interest earned on your money, instead of a 100% rate of return. Don't just think, "You know what they say about things that sound too good to be true..."

Again, performance contractors know that what I'm telling you is true. They are relying on you not believing it! It's the fast payback opportunities that help assure performance contractors that the big ticket, money maker items, will have the ROI's they guarantee.

Once again, take your administration on a tour of your facilities. Get them into the boiler room. Show them the 50-year-old boilers that have a published life expectancy of just 25 years. Get them into the equipment rooms that look more like storage rooms. Open the doors on the air-handling units to see the deplorable condition of the heating and cooling coils. Show them the steam blowing out of the condensate tank, indicating the need for steam trap repairs or replacement. Show them your 1-year and 5-year plans and budgets. Put the responsibility where it belongs – on them, not you.

I'm convinced that, following the recommendations of this book, the end cost to your facility will be significantly less, versus having a performance contract. Especially if you're working with contractors that you know, like and trust.

I also just learned of a national organization of performance contractors called the Energy Services Coalition. You can find a fairly complete listing of private energy performance contractors at their website: www.energyservicescoalition.org/

BONUS - More things that are good to know, but don't fit into the six steps

7.C – Carbon dioxide

Some time back, this whole carbon dioxide/global warming thing finally hit home for me. On the TV news, they were saying that, for the first time in the history of mankind, ships were going to be traveling between Europe and the Orient, through the Arctic Ocean! That's a little scary. It's probably past time for us to be taking action, but now is better than later.

It has taken nearly my entire life to properly comprehend the most basic lesson of high school physics. That is, matter is neither created nor destroyed, it merely changes state. Except for the meteors that didn't burn up upon entry, the earth weighs the same now as when it was first created. I used to think that, when something burned, it was gone. Now I understand that, while I may no longer be able to see it, it doesn't mean that it's gone. Instead, it simply means that it has turned into something else - mostly carbon dioxide, CO_2, and water.

Well, here is a sobering statistic regarding the CO_2 that we all generate, every winter. For every 1 billion Btu's of natural gas burned unnecessarily, it puts 117,000 pounds, or 58.5 tons, of carbon dioxide into the atmosphere!

So, how does that relate to you and me? Let's use a typical 100,000 square foot office building or school. If that facility can reduce its heating energy consumption by just 1 Btu/square foot/heating degree-day, during a fairly mild 7,500 heating degree day heating season, it will reduce the amount of carbon dioxide put into the atmosphere by 87,750 pounds, or 43.875 tons!

Elsewhere in this book, I passed on the information about the 2007 BOMA (Building Owners and Managers Association) Challenge to reduce energy consumption 30%, by 2012. On average, that's about 2 Btu/square foot/heating degree-day. I don't think my Excel spreadsheet can even calculate the tons of carbon dioxide that would reduce, if we all did our part. That is, if we all did our 30%. While the 5-year time frame of the BOMA Challenge is gone, I'm betting most of these opportunities still exist.

Many Minnesota public utilities are promoting energy savings by having rebates for boiler tune-ups, steam trap repairs and replacement, boiler replacements and more. I encourage everyone to check with their natural gas utility to learn if they have similar programs. Don't let any more rebate cycles pass you by. There is good reason to tune your boilers every year. The rebate is there to make certain that you're not going more than two years between boiler tune-ups. Some steam trap rebates are as high as 35% of the material cost.

One school we helped, reduced its heating energy consumption by 2.7 Btu/square foot/heating degree day, mostly by tuning the boilers and replacing defective steam traps. That's 2.7, times 87,750 pounds X 128,000 square feet, divided by 100,000 square feet, which equals 303,264 pounds, or over 151 tons of carbon dioxide not put into the atmosphere in one average heating season! Now, that's just from one school building. And, it was done with a 0.95 year return on the investment, even before getting utility rebates! Paybacks of less than one year really are commonplace!

BONUS - More things that are good to know, but don't fit into the six steps

7.D – Fuel cost comparison

What fuel should you burn?

Okay, you have a dual fuel, natural gas/fuel oil heating system. When times are tight, you'd like to know which fuel would save you the most money. Here is a quick comparison of natural gas and fuel oil:

Cost of natural gas	Equivalent cost of #2 fuel oil
$0.40 per Therm	$0.56 per gallon
$0.50 per Therm	$0.70 per gallon
$0.60 per Therm	$0.84 per gallon
$0.70 per Therm	$0.98 per gallon
$0.80 per Therm	$1.12 per gallon
$0.90 per Therm	$1.25 per gallon
$1.00 per Therm	$1.40 per gallon

Burning oil for an extended period of time frequently causes a decrease in heat transfer efficiency from sooting. That's because burners are tuned to burn more efficiently on the fuel used the most - natural gas. So, you might want to add $0.05 per gallon to determine the equivalent cost of #2 fuel oil, if your burners are causing any sooting. If it matters to you, the added cost of cleaning sooted boilers can add another $0.05 per gallon to the cost of #2 fuel oil.

Let's use $0.70 per Therm for the cost of natural gas as an example. If that's your current cost of natural gas, the above list would indicate that, if you can purchase #2 fuel oil for less than $0.98 a gallon, you would save money. But, if burning #2 fuel oil soots up the boiler, it may not be advantageous for you to burn #2 fuel oil, unless the replacement cost of your current supply is less than $0.88 per gallon. I'd even suggest, significantly less.

Here are some more items that should go into your decision on which fuel to use. What's the age of the oil in your tanks? Do you have any water in the bottom of your fuel tanks? Does your system have the ability to circulate the oil once in a while?

I'm not certain that oil ever gets "old", but it does seem to make a difference in how trouble free it burns. If your system circulates the oil in the tank, you should do so, a couple times a year. Keep the oil consistent throughout the tank.

There is a substance to put on the end of the stick used to check the amount of water in your oil tanks. If it turns a certain color, it means that there is water in your tanks. Water must be removed, or at the least, emulsified. Your oil provider should be able to help you with either.

If you're not using the backup supply of oil every 3 years, you should consult your oil provider. Have a sample taken and analyzed. That report will let you know if treatment should be added.

Remember, it is your responsibility to have your boilers ready to burn alternative fuels, in the event of an interruption. If you're not ready, the penalty can be substantial.

BONUS - More things that are good to know, but don't fit into the six steps

7.E Contact

Thank you, again, for the opportunity to share my years of experience with you. You can rate this book by going to: http://www.amazon.com/dp/1502764989. I'd appreciate you taking the time to rate it.

Should you want to share your comments directly with me, or have a question that I may be able to help answer, you can do so at: tom@sixstephvac.com.

If you believe this book has been of benefit to you, and your organization, I'd appreciate you telling others about your experience. Please direct them to: www.createspace.com/5040592, so they can purchase their own copies.

Best wishes!

Tom Olson

Made in the USA
Charleston, SC
01 May 2015